Carry the Fire

CARRY THE FIRE

THE CRUCIBLE OF LEADERSHIP IN THE FIRE SERVICE

KRISTOPHER BLUME

FIRE ENGINEERING™ • BOOKS •

Disclaimer

The recommendations, advice, descriptions, and methods in this book are presented solely for educational purposes. Photos are for instructional purposes only. Always wear the proper level of approved PPE when conducting training drills and operating at incidents. The author and publisher assume no liability whatsoever for any loss or damage that results from the use of any of the material in this book. Use of the material in this book is solely at the risk of the user.

Fire Engineering Books
110 S. Hartford Ave., Suite 200
Tulsa, Oklahoma 74120 USA

800.752.9764
+1.918.831.9421
info@fireengineeringbooks.com
www.FireEngineeringBooks.com

Executive Vice President: Eric Schlett
Vice President, Group Publishing: Amanda Champion
Vice President of Content Operations: Starlet Franz
Sales and Customer Service Manager: Lane Nash
Managing Editors: Diane Rothschild and David Rhodes
Production Manager: Tony Quinn
Book Development Editor: Daniel Edward Petrino
Book Designer: Robert Kern, TIPS Publishing Services, Carrboro, NC
Cover Designer: Brandon Ash

Cover photograph by Ziemba Photographic Arts

ISBN: 9781593705169

Library of Congress Cataloging-in-Publication Data Available on Request

Printed in the United States of America

2 3 4 5 28 27 26 25

For Wyatt Christian Blume

In these pages, you will find tales of courage, sacrifice, and dedication—values that define the life of a firefighter, values that I see in you every day. You have grown up with the sirens and the stories, and I have watched with pride as you've witnessed the true meaning of service and community.

May this book remind you always of where you come from and inspire you to carry forward the legacy of those who serve others above themselves. You are my pride, and this story is as much yours as it is mine.

With all my love and admiration,
Dad

CONTENTS

FOREWORD

Every now and then an author comes along who gifts our industry with an important piece of work that provides a refreshing mixture of research-based knowledge and practical application methods. Kristopher Blume is one of those authors, and this book is one of those gifts.

In the fire service, it is not uncommon for organizations to promote people into high-ranking leadership positions without providing them with any formal—or informal—training on the leadership challenges they will face in their new role. I became a deputy fire chief at the age of 41 years. Like many in an executive leadership position, at times I found myself perplexed and overwhelmed by some of the challenges that those in my position can encounter over the course of our careers. It wasn't unusual for me to be spending countless hours of my personal, off-duty time researching topics like culture development, goal setting, and strategic planning, while also trying to figure out the best ways to mentor, coach, delegate, and prioritize. Like most, I also had to deal with legal issues, organizational freelancing, and maintaining self-awareness and control while trying to develop a strong command presence and trust among my subordinates. Looking back on my personal journey, I only wish I had this book in my library to help me navigate through that labyrinth, because on the pages that follow, Kristopher Blume addresses all those issues, and so much more.

Nobody in today's modern fire service should feel as if they are carrying the weight of an organization by themselves. The best organizations are run by individuals who know how to manage, motivate, and inspire others. Leadership is not complicated for some people, but it is never simple. As human beings, we are emotionally driven, and as a result, we often make decisions based on the state of our mind at the time. This book is principle driven and presented in such a way that it can provide chief officers and organizational leaders with a blueprint on how to lead in today's complicated world.

Carry the Fire covers important topics such as leadership methodologies, brand management, legal challenges, health and wellness, hiring and retention, and relationship building. If you are ready to move forward and lead your organization, this book can provide you with the foundation you need to build a strong culture that will lift those around you—and your organization—to a better place.

Congratulations on making the choice to read this book. By doing so, you are taking one more step in becoming a better educated and more competent leader.

—Frank Viscuso, deputy fire chief (retired),
best-selling author of *Step Up and Lead*

***

I've always found it invigorating to open a new leadership book and soak up the lessons within. *Carry The Fire* is even more exciting for me as I have watched the lessons you are about to explore lived out and realized in the life of my friend Kris Blume. You are not just getting ivory tower theory here or regurgitated bumper sticker slogans. This is leadership lived out loud with all the highs and lows that will likely accompany your own journey.

Typically, a book will take a more lower shelf approach, highlighting simple messages that can be easily absorbed and applied. Others will reach higher on the shelf and attempt to address more complex ideas and the challenges of trying to lead effectively. Both approaches are necessary and can be helpful in expanding one's knowledge and skillset.

Chief Blume has taken the road less traveled with this excellent book. His outstanding insights are targeted at all levels of the leadership discipline, and you'll find a wealth of incredible knowledge to explore. Both lower on the shelf and way up high. This is a resource that will serve you long after your first reading and likely be peppered with notes and reminders that track with your leadership growth.

This is good stuff. Really good stuff that will draw you in and immediately inspire you to improve your game. Everyone leads in some way, no matter their position in an organization, and Chief Blume has written this with that reality fully in mind. You will see yourself in his words and find ways to identify the gaps in your leadership and how to address them. From the opening conversations about leadership styles, ethics and accountability, to the operational necessities of trust, temperance and planning, you're in for a master's class on leading well.

I've not read a book in the fire service that is this comprehensive and detailed. Each chapter serves as a mini book of sorts, with a depth that is spectacular in its impact, but somehow made applicable by the skill of the author. It is easy to get lost in the brilliance of the writing and the joy of anticipating putting the knowledge to use.

I have watched Chief Blume's journey with great interest and enjoyed the great successes he is having. While many fire departments are struggling with low morale, challenges in hiring and confusion of mission priorities, Kris' department is thriving. The reasons for that success, in face of so many challenges, are contained in this book. His accomplishments are not a surprise, as I have found him to be an authentic leader and man of impeccable character and skillset. This comes through so clearly in the warmth of his instruction and the passion that obviously burns behind it all.

You are in for a treat as you open these pages. This is a leadership journey of truths learned through actual time and effort. Not a theoretical mishmash, but vetted and proven strategies and tactics to help others live their dreams and the mission to be fulfilled. It's an honor to open the door to this wonderful book and you'll be a better leader, and person, on the other side of the journey. So, buckle up and get ready for an exhilarating ride.

—Capt. Mike Gagliano
Seattle (WA) Fire Department (retired)

PREFACE

Welcome to a journey of leadership in the world of the fire service, a domain where every decision can have profound implications not only for the team you lead but also for the community you serve. *Carry the Fire: The Crucible of Leadership in the Fire Service* is specifically crafted for fire service officers and chiefs who stand on the front line of crisis management and community safety.

Leadership in fire service is unlike leadership in any other profession. It demands a blend of courage, empathy, and strategic foresight, along with an unwavering commitment to safeguarding lives and property. The lessons and insights compiled in this book draw upon decades of firsthand experiences tempered by the flames—literal and metaphorical—from which I emerged with valuable wisdom to share.

Each chapter of this book addresses a core aspect of leadership within the fire service—from developing effective communication and team cohesion to managing organizational change and fostering a culture of continuous learning and improvement. I also explore the psychological and emotional competencies necessary to lead effectively under pressure, the strategic use of resources, and the pivotal role of community engagement.

In addition, this book delves into the evolving landscape of fire service leadership, reflecting on how changes in technology, society, and the environment are reshaping the challenges and opportunities you face. This is intended to provide practical guidance on how to adapt and thrive in this dynamic environment, keeping your leadership approach as resilient and responsive as the teams you command.

As you turn these pages, I invite you to reflect on your own experiences, challenge your preconceptions, and consider new perspectives. Whether you are a newly appointed officer or a veteran chief, my goal is to equip you with the insights and tools necessary to lead with distinction and impact.

I hope that this book serves as both a guide and an inspiration, helping you to elevate your leadership and thereby enhance the effectiveness, morale, and safety of your fire service team. Thank you for your dedication to this noble and critical profession. Now let's embark on this path of learning and leadership together.

—Kris Blume

INTRODUCTION

"Fired up! Fired up! Fired up!" the class of 2000–2001 yelled across the Tucson Public Safety Training Center, eager to begin our fire service careers with the Tucson Fire Department. The echo of our chant had barely left the air before I realized it was over—all 22 weeks of grueling training in the Arizona desert.

When I think back to that day, I often wonder what I would now tell the new probie firefighter I was if I could have 5 minutes with him. I imagine I would take him aside and tell him that it wasn't the end but the end of the beginning. I look at photos from that day and still have a special place in my heart for that young man: Innocent. Strong. Eyes bright. Pride emanating from his posture.

More than two decades later—through countless fires, emergency calls, births and deaths, soaring highs and devastating lows—I wouldn't change anything. Not one thing. This is the test. There is no shortcut, no fast forward. The crucible is about making every day matter and staying gritty.

"Earn it every day": these were the words of advice given to me as I retired after my first 21 years in the fire service, heading to a fire chief job in another state. I couldn't get those four words out of my head. They flipped from a maxim to a question: Had I earned it every day of the past 21 years? Had I done everything I could to make every day matter, every shift count?

I imagine every author has fantasized about a time and place where readers gather and share the benefits of reading their book. Mine would be the tailboard or kitchen table of any fire station. I hope that the content and value of *Carry the Fire* will benefit those discussions and enhance the most incredible job in the world.

The fire service is a proud, honored, and sacred profession. Our trials and tribulations, lessons, and history are built with blood, sweat, tears, and more blood. When people ask me what I do, I tell them I am a firefighter. I get the polite "Thank you for your service" response 9 times out of 10. I am ready for that one, and I immediately reply, "No, thank you for the privilege of being able to serve."

Carry the Fire intends to provide an access point for fire service leadership to gain insight and answers to the most challenging questions we encounter in the modern fire service. Most leadership books in the business and military arenas offer a tangential bend that can be related to the fire service. Still, none of them directly tackle the issues that we in the fire service encounter. With this book, that is my mission.

In this writing, I draw on the past two decades of fire service leadership and management to provide a solid foundation for current and future fire service leadership. I have muted the false bravado and platitudes. This book provides a backdrop and platform for the lessons I have learned throughout my career, driven as I pursued excellence. It wasn't always fun, it certainly wasn't easy, but I wouldn't change a moment—not any of it.

The most incredible honor in this profession is earning the firefighter title. Not fire chief or captain—those are postscripts. In the minds of your community, children, family, and elected officials, you are a firefighter. There is no greater title and honor than being recognized by your peers and others as a firefighter. In my heart, I will forever be a firefighter.

> *What you will leave behind is not what is engraved in stone monuments, but what is woven into the lives of others.*
>
> —Pericles

I hope that this book will be woven into the fire service profession, to benefit those who enter this challenging but worthy line of work and make a difference where it counts.

1 EXPLORING TODAY'S LEADERSHIP METHODOLOGIES

Leadership is the capacity to translate vision into reality.
—Warren Bennis

Not all leaders are created equal. That is a truth that those who study leadership and those who have developed themselves as leaders come to understand. By picking up this book, you've taken the first step toward genuinely understanding leadership. To do so, you must learn about the current theories, the most critical methodologies, and the best practices.

Leaders do not all follow one pathway. Over the years several theories have been developed, studied, and applied to a wide range of industries, including the fire service. From transactional leadership to servant leadership, there are a plethora of methodologies to choose from. In this chapter, we will explore every relevant one of these, including their benefits and challenges when applied to organizations. First, let's explore the traditional theories: transactional, autocratic, and hierarchal leadership.

Transactional Leadership

There are two main categories of leadership: *transactional* and *transformational*.[1] Transactional leadership focuses on the exchanges that happen between leader and follower. In transactional leadership, leaders clarify goals, set expectations, and then establish parameters for any punishments or rewards. You can probably come up with a few examples where you have encountered transactional leadership. Numerous situations and industries use this theory.

The basic factors defining transactional leadership relate to contingent rewards, active management by exception, and passive management by exception. Contingent rewards are provided once clear, established expectations and performance standards are met. Consider a sales department where the sales leaders receive bonuses if they reach a certain quota. That is an example of transactional leadership at its finest![2]

Active and passive management by exception are critical but often overlooked aspects of transactional leadership. In active management by exception, leaders take a more intentional role in overseeing the tasks at hand and intervening whenever they see fit. By contrast, passive management by exception is the opposite: they do not intervene unless problems arise and they see that their guidance is needed.[3]

There are several benefits to this leadership theory in practice. First, there are clear expectations, and the leaders themselves enjoy more success because team members know precisely what is expected of them and strive to reach this standard. Quick problem-solving is another advantage, as whenever leaders see an issue, active management by exception allows them to address these issues the moment they see the problem. Motivation is also provided through the rewards, which is why sales managers see high performers and people aiming for the top. When there's a clear standard to meet, people will be more motivated, and they will feel more in control to reach that expectation.[4]

Transactional leadership nonetheless has its downfalls. For one, the focus on compliance can be seen as a motivating tactic through external rewards. As tempting as these are, not all team members respond, and there should be a healthy dose of intrinsic motivation to ensure that team members feel fulfilled in their work. In the long term, this methodology also struggles to achieve the same success as others. Leaders who rely on transactional leadership may limit creativity and adaptability, leading to less satisfaction among their employees and even difficulty encouraging team loyalty.[5]

Autocratic Leadership

As another distinct traditional model, *autocratic* leadership allows the leader complete control over decision-making and all team activities. This is perhaps one of the most limited leadership methodologies, but it is imperative to cover it because it exists in society. Numerous examples come to mind in the various industries and situations in the world around us, from small businesses to political dictatorships.

Autocratic leaders are generally defined by a few factors. First is their centralized decision-making control: They do not allow anyone else to take part in their decisions. Period. Even if part of a team, autocratic leaders are responsible for making choices. They also tend to be highly dictatorial, meaning they do not relinquish control to their subordinates. Their team must heed their commands, even if they do not feel it is the best choice for their organization or fire department.[6] This leaves employees without much autonomy—if any—and this theory has several challenges when used in practice.

Despite the precise direction, quick decisions, and streamlined processes that autocratic leadership allows, personnel issues arise since team members are entirely disregarded. They are likely not as involved, leading to increases in poor employee morale and motivation. Think about it: What employee would be motivated to work for a leader who does not care about what their team contributes? This can ultimately lead to less employee retention—as you might have guessed.[7] Overall, this is a narrow leadership philosophy and one I don't recommend implementing in your organization.

With that in mind, you might wonder why this was brought up. After all, if this is a book about leadership in the fire service, why talk about the least effective leadership theories?[8] The short answer is this: to learn what *not* to do. Autocratic leadership is also a form of hierarchical leadership, which has some benefits and similar challenges.

Hierarchical Leadership

Much like autocratic leadership, *hierarchical* leadership is a command-and-control leadership style. The leadership has the same power as in autocratic leadership, yet with more emphasis on the chain of command. When authority and hierarchy take precedence, each level of the hierarchy determines who has power and to what degree.[9] Leaders at the top will have more and delegate to those at the bottom, contributing to an overall leadership structure that can become rigid and formalized.

Most organizations today have some elements of hierarchical leadership owing to the roles and the levels of management that people see. For instance, consider Coca-Cola. They have their owner, who has the most power; then they have executives, department managers, and managers; and finally they have associates. Each of these groups has a place in the chain of command. Similar situations arise in fire stations; everyone defers to the fire chief or the senior members and roles within the station.

With hierarchical leadership, there are clear lines of communication, power, and formalized top-down communication through which directives and delegations are passed. Unlike autocratic leadership, all people have a role to play and some degree of influence—big or small. This can be beneficial, by instilling clear structure and order and allowing efficiency in operations. However, there are a few issues regarding knowing who to turn to when there are not clear guidelines and transparent chains of command. When the roles are defined and delegated, people can take ownership of their accountability and consistency.

This structure comes with similar challenges owing to its rigidity. The roles and the delegation of tasks can make creativity nonexistent and limit how involved the employees are in their fire department. Those lowest in the command structure may also struggle with poor motivation, battle communication challenges, or feel micromanaged. With traditional leadership theories such as this one, it can be difficult for employees to remain motivated, yet an understanding of these theories also helps to explain the draw and the transformative nature of other theories that break free from the expectations of the past to pave the way for innovative futures.

Transformational Leadership

Most leadership theories are characterized as either transformational or transactional. By contrast with transactional leaders, transformational leaders are less focused on metrics and exchanging punishments and rewards. These leaders are revolutionary for their inspiration, motivation, and innovation.[10]

You can probably come up with several examples of transformational leaders: a coach from your childhood, a boss who invested time in inspiring and motivating you, or even a fictional character from a movie. Transformational leaders are all around you, and there's a reason why many remain memorable, even years after you have encountered them: Transformational leaders encourage their team members to grow, and they challenge the status quo.

In transformational leadership, the rigid structure that characterized autocratic and hierarchical leadership is relaxed. Transformational leaders constantly look to the future and question tradition. They are best known for their ability to inspire motivation. They take their employees into account—individually—to ensure that team members grow because of their mentorship. These leaders also serve as role models. As we will discuss in chapter 6 (see "Integrity and Its Impact on Workers"), these leaders are likely also to possess integrity—one of the most critical factors to ensure that your team members not only follow you but want to become like you as leaders themselves.[11]

The benefits of this leadership philosophy are numerous. Leaders who practice this methodology improve employee engagement, increase employee satisfaction, and enhance their organization's creativity and innovation. As mentors, they develop leaders within their organizations, creating legacies and opening opportunities for organizations to become more robust with each new generation of managers.

This leadership style has challenges, too. There will always be team members who resist change, making the leader less effective. This can also be demanding on the leaders in both time and energy. Not all leaders can master this style, but many aspire to. Moreover, there are also concerns if the leader struggles to remain ethical in their guidance.[12] By contrast, if leaders uphold strong morals, there's no question that they can improve and innovate within their organizations.

John C. Maxwell, one of the foremost speakers and writers on leadership, believes in the power of transformative leaders. In *Change Your World: How Anyone, Anywhere Can Make a Difference*, Maxwell and coauthor Rob Hoskins state, "If you're leading a group of people, you have to realize the team and the transformation they are creating are essential to realizing success."[13]

Similarly, transformational leaders look at their teams, paving the way to success unparalleled, as compared with traditional theories. Yet, this theory is not the only one that has such beneficial impacts on teams and organizations.

Servant Leadership

Do you love your job? And do you truly serve the work? I have found that *servant* leadership is one of the most common leadership theories throughout the fire service, mainly because of the degree of collaboration and fulfillment that comes from what we do as firefighters. Our job is to serve others, to fulfill the community's needs, and to protect others. As public servants, we are—by nature—looking at how we help others first, committing our energies to the betterment of our world and community members.[14]

Like transformational leadership, servant leadership is an impactful, people-focused initiative that organizations have implemented worldwide. This leadership style values service, empathy, and collaboration, but most important, it emphasizes the people involved in the organization. To put it plainly, servant leaders serve others.

While many people are driven by the monetary gain that comes with their jobs, firefighters—and anyone for that matter—should be motivated not by money but instead by the intrinsic value of their heroic acts. In the fire service, there is undoubtedly intrinsic value in serving the public as a protector of the community. For many of us, our ultimate motivation is that we love what we do, and we do it for the community. Not all fire chiefs are the same, but remembering that our ultimate motivation is about the people we serve is part of why firefighters align most with servant leadership, a concept that has been around for over 50 years now.[15]

The first mention of servant leadership was in 1970 by Robert K. Greenleaf, who discussed how leaders can put the needs of others ahead of their own interests.[16] Think of John F. Kennedy's famous line: "Ask not what your country can do for you—ask what you can do for your country."[17] Servant leaders ask what they can do for others instead of what their teams can do for them.

Many people also see servant leadership through the lens of the age-old belief that you will never work a day if you love what you do. For public safety, it is all about fulfillment in serving others and serving the work, which should be—and hopefully is—a labor of love.[18]

Servant leaders have defining characteristics that set them apart from others. First, they prioritize listening and understanding each person's needs and perspectives. They seek to understand and guide first, rather than delegate. They also have empathy and compassion for others. Their level of support for their team members is unparalleled within almost any other leadership philosophy. Furthermore, servant leaders practice self-awareness and promote healing to ensure a supportive, positive environment focused on productive and transformative change.[19]

Servant leaders typically have a clear vision and a deep understanding of their team. They can provide direction and make decisions that benefit every team member. They also prioritize responsibility for their team's well-being. They practice complete and total accountability—that is, what happens to their team is their responsibility and nothing less.[20]

Servant leadership has numerous benefits and is commonly associated with its positive attributes. These leaders have higher employee satisfaction and stronger teams. Their teams also succeed, demonstrate improved performance, and enjoy a positive organizational culture. With a leader like this, it is impossible to imagine feeling otherwise.[21]

Servant leadership can also empower each firefighter—and the entire staff—to continuously improve their skill set and education. When you seek to serve the work, you seek to perform beyond your expectations—you reach for the stars knowing that even falling short of your goals still benefits your stakeholders. I truly believe that becoming an influential leader in the fire service is about going beyond mere proficiency and becoming a pursuer of excellence in all that you do. Continuous improvement—for yourself and your team—is what sets servant leaders apart from others and what amasses benefits for them and their organization in the long term.[22]

Much like transformative leadership, servant leadership can be time consuming, and it can also receive a lot of pushback. Team members might desire more delegation and less counseling from their leadership. To succeed, team leaders who practice servant leadership should avoid becoming overly permissive and allow their team to fully direct their choices.[23] Servant leaders serve their team, but to ensure that their team succeeds, they must also lead.

Authentic Leadership

Authentic leadership is a unique methodology, as it demonstrates the leader and their personality the most out of all these leadership theories. An authentic leader emphasizes their self-awareness, focuses on transparency, and stays true to their values. This is the theory that leaders are true to who they are and, because of that, most likely to inspire trust and engagement among their team.[24]

Authentic leaders recognize what they excel at versus what needs improvement. These leaders are transparent and do not try to hide mistakes from their peers. They are also relational and empathetic to others. They want to foster a sense of trust within their team. Furthermore, they are consistent and resilient, two qualities that are essential to be a strong leader in any industry or organization.[25]

With this leadership style, teams often enjoy greater trust in their leadership and experience more engagement. This may be directly rooted in their leadership cultivating a positive organizational culture. This can—and usually does—lead to employee development and improved performance. However, challenges may arise. In some cases, these leaders can be perceived as inauthentic and insincere, which can damage their influence and the degree of trust others have in them.[26]

Authentic leaders might also require more time and energy—as is generally the case for any transformational leadership style—so this can be difficult to implement. Finally, not everyone will be open to change, but with the right mindset and the right approach, this theory could genuinely instill transformation in any organization and team environment.[27]

Situational Leadership

Situational leadership is precisely as it sounds, with its emphasis being on adapting one's leadership tactics to the situation presented. This means that there is no one-size-fits-all approach, but rather the approach will be tailored to each unique challenge that the organization might face.[28]

There are generally four leadership styles that are part of this methodology. From directing to supporting, these are the tactics that a leader must learn and implement with their teams.

Directing

As the name implies, *directing* is generally defined as leading by applying more directive emphasis than supportive behavior. In this leadership style, the leader provides specific instructions and makes decisions for the team. There is generally more supervision in these situations with the followers being guided by their leader. For situations like crises and other high-commitment expectations, this is the best tactic to use.[29]

Coaching

With the same level of directing but equal levels of support, the *coaching* tactic helps teams to succeed under situational leadership. This is when leaders guide and support their teams while giving them a chance to participate in the efforts. This works best when team members have a greater understanding of what is required and the skills to do so but struggle to commit. A great example would be when a project is within the employees' skill set but their commitment is low, with the team leader responding by taking a more active role to ensure that the task is completed proficiently.[30]

Supporting

With less emphasis on directing, *supporting* mimics aspects of servant leadership. The leader prioritizes supporting their team members, as well as encouraging them. Team members are also given more autonomy than they would in a situation where coaching is used instead. When team members have a greater degree of commitment and high to moderate competence in the assigned task, this is the best approach to use. One example of this would be when executives are available for support and have questions when a marketing team is curating their presentation, but the team members are driven and competent enough to deliver the expected results.[31]

Delegating

Finally, *delegating* is when there are few directives from leadership and little support. This is a more passive role where the leader empowers their team members to take ownership of the work they are completing. This is great for teams that are highly committed and proficient in their tasks, as this can be a more relaxed tactic that ensures that the team succeeds with little intervention from the boss or management.[32]

***

One of the most significant benefits of situational leadership—as evident in the four main dimensions that one could apply—is the excellent degree of flexibility that leaders can exercise. It is clear how this could benefit nearly any organization. In a crisis mode, a situational leader could call upon directing to ensure the tasks are done. For more routine and manageable tasks, a fire chief, for example, could delegate the work to the station and allow them to take leadership of what has been asked of them. The options are truly endless—and the impact of this leadership style has no bounds.[33]

Adaptive Leadership

There are several more essential leadership styles to round out the discussion of leadership methodologies and current best practices: adaptive leadership, ethical leadership, meta-leadership, and purpose-driven leadership.

Much like situational leadership, *adaptive* leadership involves a degree of flexibility that allows leaders to guide their teams in the right direction and through all transformative eras. Leaders use this style when they need to navigate unprecedented changes. From technological shifts to massive staff changes, leaders must be prepared for it all, and this leadership style takes the need for flexibility into account.[34] Instead of trying to be the person who knows how to navigate every situation, adaptive leaders focus on curating the right environment to empower their team for significant change.[35]

There are a few critical components to adaptive leadership. First, there is the *diagnosis*. Much like a doctor, a leader must understand what is happening and the complexity of the team. For instance, a team might be struggling with losing a critical leader for one of their departments. Adaptive leaders gather a comprehensive understanding of what is happening, understand the nature of the situation, and then figure out how to respond.[36]

Next, leaders will implement an *intervention*. This is meant to engage their stakeholders and employees to encourage problem-solving, including using alternatives that might be outside the box. The intervention is then applied to address these issues and to overcome the challenge.

Finally, *learning* is the last key component of this methodology, as leaders create continuous education opportunities and prioritize the ability of their team to reflect on their experiences. They want their team members to understand and learn and their culture to be curious and adaptive. More important, they also believe that failure is a great teacher.[37]

There are several benefits to this methodology. While seemingly like situational leadership, adaptive leadership also encourages innovation and resilience. Adaptive leaders encourage their teams to grow, evolve, and overcome challenges innovatively. These leaders also encourage individuals to view their setbacks and failures as chances to improve. Adaptive leaders tend to influence their teams to become adaptable to all situations that might arise. Under adaptive leadership, teams transform, but they also work proactively to prepare for any situation that they might experience together.[38]

Ethical Leadership

Leaders need to have high moral standards, which explains why *ethical* leadership is a critical current methodology that has been in the industry for several years. This approach to leadership emphasizes the moral responsibility, integrity, and careful decision-making of the leader and the team. These leaders aim for accomplishments but also prioritize values that ensure that they benefit and respect their various stakeholders.[39] These leaders set a firm example of how to act responsibly, foster trust, and ensure credibility.

Ethical leaders play an essential role in developing organizational culture and implementing critical moral codes. Most important, these leaders set the tone for the rest of the team.[40] Their example demonstrates precisely what is expected, what should be practiced, and what the organization likely should strive toward. With every decision that aligns with their ethics, they are creating an organization-wide standard that cannot be ignored.

Furthermore, ethical leadership also establishes precise values. Without a moral leader, many teams might find it challenging to define their team values. However, an ethical leader will exemplify those values, providing a clear example for the team to follow. Thus, team members align their values over time.[41]

Ethical leaders also remain valuable to organizations because they cultivate open communication. An established framework of ethical conduct can foster an environment where open communication is valued and encouraged, in which employees can come forward to voice their concerns and ask questions.

As an example where ethical leadership is needed, let's consider a bank where misconduct is rampant. There would be little opportunity or acceptance of those who call out this behavior if leadership did not set an ethical example. In an ethical leadership setting, though, the leader—leading by example—would ensure that people felt comfortable coming forward to report violations that could otherwise harm the brand name and their reputation in the public eye. Similarly, ethical leadership could keep fire stations in line when members suspect unethical conduct.[42]

One reason why ethical leadership is essential in today's world is the demand for corporate social responsibility. Today, every organization in every industry must prioritize environmental, ethical, and social responsibility. The community and the stakeholders have thankfully been endowed with more importance in the way that organizations decide what to do and how to act.[43]

One example of social responsibility is given by the shoe company Toms. They set the tone in their industry by being one of the first organizations to make a pair of shoes and send it to a child in need for every pair of shoes they sell. This ethical conduct is—above all—a shining example of corporate social responsibility.

Another excellent example of social responsibility is given by Jamie Dimon, leader of JPMorgan Chase. His contributions to K-12 education in the United States in recent years highlight the ethics and moral code of his organization and the society it serves.

Ultimately, the most ethical leaders are those who make a difference for their organizations, stakeholders, and employees. A strong sense of employee trust and organizational reputation is fostered when ethical leadership is prioritized. Much like the other transformative leadership styles, you can imagine what this does for any organization: It encourages trust, communication, commitment, and loyalty.

Fire stations with ethical leaders are seen more positively in the public eye. This not only ensures easy brand management but also can lead customers and stakeholders to have more favorable opinions of the organization itself.[44] As you can see, this leadership style has great applicability—and even more importance in today's world, where ethics are part of the fabric of every organization.

Meta-Leadership

The concept of *meta-leadership* was first introduced in *You're It*, by Marcus, McNulty, Henderson, and Dorn.[45] This concept is essentially a framework for three important, connected components: the person, the situation, and the connectivity. By understanding all three of these components, leaders can excel and guide their teams across organizational boundaries, especially during the most challenging times for their teams.

Next, let's dive deeper into each component and why it is essential to this framework.

The Person

This component entails the individual leaders themselves. This is about understanding who the leader is, including their strengths and weaknesses.

As part of this framework, leaders must be self-aware, be committed to values that are ethical and beneficial, and exhibit emotional intelligence.[46] If any of these factors is missing, it can lead to challenges for the fire station and how it operates. Additionally, leaders also must be emotionally mature and empathetic. Leaders cannot be distant, and they cannot expect to lead without compassion. In meta-leadership, leaders are an essential component to the overall success of their team, overcoming their challenges to succeed in the end.

The Situation

Meta-leadership is about more than just the leader; it is also about the situation or the challenge at hand. This component entails understanding the broader situation and the complexities of the organization, including their industry.[47]

Leaders not only have to be strong role models for their teams but also must be able to navigate diverse stakeholders. Furthermore, they must understand how to balance competing interests and address financial, technical, and operational challenges.

The Connectivity

Finally, in tandem with the person (i.e., leader) and the situation, meta-leadership considers the situation's complexity. This entails how to connect and collaborate with others, often across boundaries, to achieve goals.

Leaders who utilize meta-leadership must be able to build bridges, transcend conflict, and collaborate with others to achieve their goals.[48] If the connectivity is missing, then the leader cannot adequately solve the situation and the people fail to understand their collective roles. Since meta-leadership can fail in these situations, connectivity remains an essential aspect of the leadership framework and its overall success rate.

***

There are many benefits to meta-leadership—when applied well, of course! Leaders who apply this successfully can enhance team collaboration, improve decision-making skills, and provide practical, prepared responses for crises. Moreover, such leaders tend to provide adaptive solutions that are also sustainable, leading to lasting change or balance during the most troubling of times for the team.[49]

However, this leadership framework is also challenging because of conflicting interests and power dynamics, which can make it difficult to allocate the resources appropriately. It is also demanding of time and energy. With any leadership theory, not everyone will buy into the philosophy or framework, so it is important to weigh the possibility that collaborative relationships might not be achieved. Nonetheless, meta-leadership remains a critical component of contemporary leadership frameworks.

The Power of Purpose-Driven Leadership

There are various theories, frameworks, and contemporary examples of what leadership is and is not. However, one leading theory or principle guides every leader, regardless of their chosen framework: their *why*. That is the basis of purpose-driven leadership.

Simon Sinek, an experienced leadership scholar, gave a TED Talk about the importance of leading with a central purpose. He stated that all leaders should "start with why"—that is, find their purpose and then be able to let that drive them to success. He even wrote a book on the topic, *Start with Why: How Great Leaders Inspire Everyone to Take Action*, where he explains how the purpose and the mission ultimately drive every leader to their designated level of success for them and their organization.[50]

Across diverse leadership theories, the common theme is that every leader can have their own framework and methodologies—or even multiple—but every leader must have a purpose. By understanding the *why*, according to Sinek, leaders can then try to identify the *what* and the *how* of their work for maximum efficiency.[51] The *what* are tangible products, services, or even actions that leaders can pursue or complete with their teams. The *how* comprises the strategies and processes that ensure their efforts succeed. However, Sinek stresses that without a clear *why*, the *what* and the *how* do not matter and cannot be identified or achieved.[52]

Purpose-driven leadership is arguably what every leader should strive for. Every leader has a reason why they delegate their tasks or why they need to complete their objectives. By incorporating the tenets of Sinek's thought leadership, starting with *why*, you can become a purpose-driven leader with authentic communication and emotional engagement. You can also inspire innovation as your higher purpose.

Suppose you seek to make a big difference in your organization. In that case, consider purpose-driven leadership as one of the most important aspects of what makes you a strong leader. Whether in the fire service or any other industry, what you do and how you define your mission will ultimately define your team and your success or even influence it. With purpose-driven leadership, your team may excel to new heights and innovate, develop, and accelerate their successes.

Final Thoughts on Leadership

This deep dive into leadership theories and philosophies is critical to understand—and even more important to know how to apply in our industry. Regardless of whether your skill set is more traditional or transformative in its scope when you lead, you have the power to make monumental changes to improve your organization and align your team with your goals. The various methodologies and frameworks presented here are intended as food for thought as you progress through this book.

Regardless of your leadership theory, one crucial fact is necessary: you must remain effective. In the fire service, your leadership is paramount, and how you lead is how your team serves. This book aims to inspire you to carry the fire to be an effective leader, not only for yourself but for the impact that you can create within your team and your community.

Chapter 1 Review Questions

1. What are the two main categories of leadership?
   a. Open and closed leadership
   b. Transformative and formative leadership
   c. Transactional and transformational leadership
   d. Strategic and operational leadership
   e. Participative and authoritative leadership

2. What type of leader sets expectations and parameters for punishments or rewards?
   a. Servant leader
   b. Hierarchical leader
   c. Transactional leader
   d. Transformational leader
   e. Autocratic leader

3. Which of the following is a benefit of successfully applying meta-leadership?
   a. Increase in power dynamics
   b. Decrease in decision-making skills
   c. Building sustainable collaborations
   d. Reduction in team collaboration
   e. Lack of adaptive solutions

4. Why is ethical leadership important for organizations in today's world?
   a. To prioritize profit over social responsibility
   b. To focus solely on personal gain
   c. To set an example of corporate social irresponsibility
   d. To emphasize corporate social responsibility

5. Which of the following are the three important components of the meta-leadership framework?
   a. The organization, the resources, and the challenges
   b. The person, the situation, and the connectivity
   c. The power, the control, and the communication
   d. The vision, the mission, and the values

6. Passive management by exception in transactional leadership involves addressing issues immediately.
   a. True
   b. False

7. Sinek's thought leadership emphasizes starting with why to become a purpose-driven leader with authentic communication and emotional engagement.
   a. True
   b. False

8. It is not important to know how to apply leadership theories and philosophies.
   a. True
   b. False

9. Under purpose-driven leadership, the *why* is the central driving force behind leadership success.
   a. True
   b. False

10. The text focuses on transformational leadership theories, rather than traditional leadership theories.
    a. True
    b. False

11. What qualities make an authentic leader stand out in an organization?

12. How can servant leadership empower firefighters and staff to seek improvement in their skill set?

13. Why do ethical leaders play a crucial role in setting organizational standards?

14. Why is it important for a servant leader to have empathy and compassion for others?

15. What are some potential benefits of successfully applying the meta-leadership framework?

2 ORGANIZATIONAL CULTURE

You've heard the word *culture* before, I'm sure—and not just for describing societies. In general, culture comprises the learned behavior patterns of people. It includes what people say, feel, think, do, and value. In a professional setting, it involves beliefs, pervasive values, and attitudes that influence the profession.

You might have heard the term *organizational culture* as well—especially as it's a buzzword in the corporate world. It encompasses the organization's values, expectations, and ethics. It's also something that employees are expected to adapt to. At its core, it makes or breaks the organization. If your organizational culture is poor or nonexistent, your organization will suffer and may struggle to achieve its goals for the future.

Every organization has a culture, even if never formally shaped by its leadership. A solid organizational culture is critical because it determines the direction your team will go. It also allows the organization to expand and reach newer heights, by increasing productivity, decreasing turnover, and promoting employee engagement. Moreover, it can motivate employees to become better versions of themselves.

If your organizational culture is positive and supportive, then your operations will similarly be positive in any industry—including fire and emergency services. If you have ever worked as a firefighter or plan to, this won't be news. The collective culture of fire and emergency services is deep rooted and solid. Moreover, traditions within our society heavily influence it, with a visible legacy of strong service provided to the community.

A Deeper Look into the Culture of Fire and Emergency Services

Culture is a guiding light for how members are supposed to act. Fire and emergency services require a team effort that prioritizes quick problem-solving in dangerous situations, often relying on the trust and effectiveness of every agency member. New recruits abide by the organizational culture to fit into their new role quickly and easily. Recruits are motivated—even inspired—by a deep desire for teamwork, clear communication, and trust among their coworkers. However, positive organizational culture is not simple to curate, nor is it easy to maintain as an ideal. Indeed, as we will discuss, certain aspects of the culture have become toxic and warrant a change.

The culture of fire and emergency services influences behavior in subtle ways, such as deciding whose responsibility it is to clean the bathroom and check the rigs. Moreover, the traditions of the fire service are time honored, as you can see reflected in the culture of a fire station or training academy: The senior members are highly respected, and the junior ones look up to them. Senior members have a more distinguished seat at the table; they also have different work assignments than the rest of the firefighters. In many ways, these differences do not impact the productivity of the team environment. Yet, a delicate balance must be struck between honoring senior members and ensuring collaboration to achieve all the organization's objectives.

This tradition is why heroism is openly displayed and embraced within the profession. It stems from the selfless way firefighters operate and how the culture influences them to become influential, strong leaders for up-and-coming department members. Firefighters must jump into burning buildings and ensure that others are brought to safety with minimal regard for themselves. Senior members are presented with so much more respect because they have endured more, have more experience, and are a shining example of the successes and progress of the fire department. As compared with other industries, the fire service has a unique organizational culture, as there is not only a hierarchy of respect but also a greater purpose expected of each firefighter on the team.

The culture of the fire service also ensures that new recruits know that the job serves a greater purpose. Many people go into a career to make something for themselves or to simply become richer. However, that's not what firefighters do: They know they aren't in it for the money. In fact, research shows that the average salary of a firefighter is just $45,000 per year,[1] which is barely enough to afford a comfortable living in many regions. Firefighters are there because they genuinely want to help, serve, and support the public. They are there because of the opportunity to make a difference in their communities.

Therefore, the mission statement of fire and emergency services reinforces the idea that members work for the community and are willing to accept risk, even if it means compromising their lives. Fire service culture prioritizes selflessness and the potential sacrifice that each member makes for others, because their duty—above all else—is to protect the public, even if they must be in the line of fire to do so.

More important—again tied to the mission of fire services across the nation—safety and continuous development represent another essential element of fire service culture.[2] From learning from the leaders of the department to engaging in routine safety training, fire departments nationally prioritize an organizational culture that values safety, education, and continuous improvement.[3] If not for these values, it would be challenging to ensure that fire departments consistently have the resources, mindset, and training to ensure that their departments effectively support and serve their communities. Without safety and continuous development, the fire service's organizational culture would differ significantly.

While they might not be harmful at face value, the structures of traditional hierarchies are subtly programming the mind to think and act a certain way. Many people in fire service organizations can and do use this to their advantage. They recognize the values of integrity, courage, and dedication, seeing these as the ideal criteria to guide their actions on and off duty.

In addition, we must remember that peer pressure exists, even in professional settings, and can be used to negatively influence incoming recruits. For example, plenty of adults spread rumors to exert power over their peers and increase their own influence. To combat negative peer pressure and related threats to your organizational culture, implement and foster a culture that is mission driven.

Mission-Driven Culture

One of the greater organizational cultures you can foster is one that is mission driven. Most fire departments in the United States use a hierarchical structure for command and control. Through this structure, the authority creates and enforces policies and protocols that guide employees during crises. Firefighters and officers in this organizational structure follow strict rules and commands given by their superiors.

As time passes and people collaborate even more, alternative leadership methods are constantly being explored. One of the most popular methods is *mission-driven culture* (MDC), which I discussed in *Fire Engineering* in 2022 after learning about it from San Diego Fire Department's fire chief, Brian Fennessy. MDC decentralizes the decisions: the organization is guided by the leader and their intentions while establishing accountability and setting clear expectations so that staff members make intelligent decisions. This is unlike traditional emergency management tactics or strategies because subordinates do not have to obey strict commands and policies but instead improve the potential for outcomes through the deviation from such rules—without consequences—

when they do not apply or serve the organization's goals.[4] Chief Fennessy's department, where the leadership ethos has become woven into the fabric of their own organizational culture, exemplifies this idea.

One essential aspect of MDC is implicit within the name itself: The focus is on each mission, allowing people more freedom to make critical decisions in challenging, unexpected situations. In other words, MDC is driven by the individual goals and mission of the entire organization. This is what firefighters work for; it is part of their daily work and how they structure their tasks. One of the most significant features of MDC, as compared with other leadership theories (as explored in chapter 1), is a move away from rote processes, instead prioritizing successful operational outcomes.[5]

In an MDC, everyone has a role. MDC has a strong foundation of core values and practices to integrate throughout your organization. Five core values espouse this philosophy:

- Service to promote the common good
- A high state of trust
- The pursuit of truth, function, and form as defined by the result
- Individual initiative
- Continuous improvement

These core values must be the foundation of any implementation of this philosophy in a fire department.[6]

There are several assumptions when MDC is applied appropriately. First, MDC is implemented with the assumption of *uncertainty*, meaning that every decision will be imperfect and must be made according to the crisis. A decision made at the moment during a crisis will always be flawed. The second assumption is *friction*, meaning the individual closest to the challenge will make the best decisions. Ultimately, what this does is take away the danger of making decisions and ensures that those who are directly facing this challenge will make the best, circumstantial, flexible decision for the moment. The third assumption is *danger*, meaning that even a well-trained and highly experienced operator will take several precautions to prevent injuries and deaths.[7]

In emergency work, no amount of education will prepare you for the challenges you face, as Heath shares in *Upstream: The Quest to Solve Problems before They Happen*. No matter who makes the decisions, there is always risk involved here, but that doesn't mean precautions cannot and should not be taken by all.[8] These assumptions illustrate how complex and challenging emergencies can be and how there is no one set of rules that fits all situations. Thus, flexibility is key to provide significant benefits when MDC is applied appropriately.[9]

The Benefits of MDC

There are several benefits to MDC. First—and one of the most important—is *focus* on the operation at hand. There is an encouragement that further illustrates a clear mission and can further inspire engagement and boost performance in this team environment. Firefighters following this philosophy will see more loyalty within their team.[10] That is because there's a common purpose with a common cause. When the job is done successfully and individually, it is no wonder that retention increases because job satisfaction is higher than ever before.[11]

Next, MDC ensures *strategic alignment.* This helps leaders to balance their priorities and establish what is necessary within the organization. Leaders can then focus on the most critical aspects of their jobs, rather than having to focus on decision-making and accountability—or having to focus on absolutely everything, even the frivolous aspects. Ultimately, this allows leaders to focus on what grows and benefits their organization rather than everything at once.[12]

Third, being focused on the mission allows for *clarity* in judgment, which can guide leaders to the best decision. Finally, as expected, operational outcomes are appropriately measured, and performance is appropriately recognized, making the ability for feedback and employee development more efficient than possible with other theories.[13]

One of the most important factors that comes with this theory is increasing the organization's accountability, speed, and balance. This has many benefits, especially in maximizing employee engagement, which can drive crucial organizational performance and increase accountability for all roles. Furthermore, autonomy and the removal of management policies that limit employees can enhance positive results across the entire organization, including in critical lifesaving actions.[14]

MDC can optimize how leaders balance safety, effectiveness, and efficiency to deliver the service needed for the community. When a culture of permission is avoided, more risks are taken; however, individuals can take opportunities that make a difference, as opposed to breaking rules and being punished, and this can ultimately create better outcomes.[15] This methodology encourages quick thinking and even quicker actions, which is essential in fire service.

The application of MDC provides numerous additional benefits. These include a decrease in grievances, an increase in retention, and even an increase in the quality of risk decisions. In turn, this leads to heightened trust, fewer surprises, and overall higher support throughout the organization, on and off the job.[16]

How to Develop MDC

As you can imagine, building an MDC is not easy, but you must consider this necessary. Cultural change can be difficult, so leadership needs to decide to institute this culture for the team. Chief Fennessy, who promotes this methodology, believes in personally investing in this change. This creates further trust and reassurance and ensures that leaders can allow their team to assume the role of leader and make decisions in times of crisis; this can ultimately create further strength, further validity in their judgments, and further empowerment of the overall team.[17]

Importantly, though, Fennessy cautions that legal counsel is necessary when making any change to the culture. Communication in these climates can be ensured to eliminate any potential problems, whether administrative or operational, that can come with these changes. When legal counsel are available, staff members can see and avail themselves of this additional support in situations requiring difficult calls. Through legal counsel, the boundaries between good and bad—or outright wrong—decisions can be articulated and made further apparent to all members of the team, including the leadership.[18]

The Leader's Role in Supporting an MDC Environment

The leader's role in the development of MDC is essential to the organization's overall success. The leader's intent is an essential aspect of MDC; therefore, the leader must lead their organization to make the changes that they desire.[19] More power is delegated to subordinates in certain situations, but this does not take away from the power of the leadership and their necessary role. For example, mission-driven executive leaders have an essential task of administering quantitative accountability in the organization, as well as embodying the needs of those that they lead, as demonstrated by individuals such as Chief Fennessy.[20] This means that individuals in an MDC are responsible for both those they serve and the employees themselves.

In an MDC, a leader's role is of paramount importance and will be focused on communicating the task, purpose, and end state of any mission or assignment being delivered. Leaders are also expected to allocate the resources needed for the operation or the assignment. This is ultimately the extent of the leader's role, as it is up to the subordinate to decide how they complete this mission following these directives. Without input from either leaders or line personnel, firefighters within all ranks of the organization are expected to act, following the organization's mission as it has been integrated into their culture.[21] In turn, MDC focuses on training all subordinates and officers to use their judgment regarding the intent of the mission, rather than obeying rules, orders, and policies to the letter. This does not mean that members are always allowed to disobey orders, but given chaotic circumstances and the dynamic of the situation, mission accomplishment may require the disobedience of the team members to get the job done.[22]

There are five crucial questions that any mission-driven leadership needs to ask to ensure their culture is as influential as possible:

1. *Are you well informed in keeping critical audiences informed?* This means that specific delegations and effective communication are needed to ensure that everybody is appropriately aware of what is going on.[23]
2. *Are you allowing calculated risks?* A fundamental benefit of MDC is that employees are concerned with making a difference—including taking risks to do so—rather than scared into indecision or adhering to policy first. This allowance for calculated risks is essential, as it increases the autonomy of every individual involved.[24]
3. *Can you effectively lead while staying out of the way?* Being a leader who can get out of the way and allow others to take charge means that you maintain your responsibilities and encourage the independence of each team member.
4. *Can you predict the future?* This relates to the implementation of programs to ensure that in future similar incidents are prevented or at least managed until team members arrive.[25]
5. *Can you do all of this with compassion?* Compassion is essential in an MDC. This yields psychological and physical benefits: lowering elevated heart rates, switching off a body's fight-or-flight threat responses, and increasing your emotional connection with your subordinates.

By asking these five questions, individuals can mitigate the difficulties that come with decision-making during a time of crisis, as well as fully understand their role and how beneficial an MDC truly is.[26]

Embracing Ownership and Organizational Accountability

Implementing MDC in the fire service requires accountability and ownership. This can be achieved in various ways. For example, line staff could be required to explain their actions to the leader. This means that they stand and explain themselves no matter what happens. It does not matter if it is right or wrong, but rather that they are accountable for their choices.

This moves away from traditional modes of management that require leadership to take accountability. Leaders such as fire chiefs are not expected to take the fall entirely for decisions with bad outcomes, but instead the accountability is dispersed among those who make the decisions themselves. This increases responsibility and connectivity across the entire organization, as well as efficiency.[27]

Achieving and Moving the Organizational Mission Forward

MDCs can help to move the organizational mission forward at the hierarchical level in fire departments. Specifically, firefighters are given the chance and scope to foster successful outcomes in their mission. In turn, this increases their job satisfaction.[28]

Additionally, MDC advances the organizational mission, allowing individuals to become more empowered in their responsibilities—with freedom to make choices and enjoy a truly satisfying and successful job. Firefighters may feel more fulfilled, capable, and willing to support the mission and operational requests. They may be allowed more opportunity to focus on risks and tasks rather than red tape, purpose rather than rules, and even end state over potential trouble for breaking a rule.[29]

Finally, MDC can positively impact the ability of fire chiefs to accomplish the organizational mission. This alleviates burden and risk factors in decision-making while supporting circumstantial decisions that are balanced and appropriate.[30]

As you can tell, an MDC within any fire department organization is one granting individuals autonomy and authority to make decisions during crises or challenging and ambiguous situations, increasing accountability and job satisfaction. The benefits of an MDC are numerous, but first and foremost, it increases the ability of firefighters to save lives and complete their jobs successfully.

Implementing MDC is tedious and time consuming, but all together cooperation, open communication, provision of legal counsel, and increased trust can create a more prosperous, integrated system.[31] To succeed, cooperation among all leaders is imperative. This is an effective system that should be implemented by all fire departments across the United States to increase the efficiency of firefighters' services and further provide value to the communities they serve.

Current Transformations of Fire Service Culture

Before we move on from organizational culture, let's explore current transformations of fire service organizational culture and how this is connected to change management. Change begins at the top, as employees are more likely to embrace the same habits if they see their leader care about the right thing and honoring the organization's profitability. The resulting change in the organization's environment will be noticeable, satisfying the department or organization's stakeholders and attracting new professionals who appreciate ethics.

Considering this, leaders must recognize that they shape the fire department's culture and principles. The organization's culture is defined by its decisions, the management approaches it employs, and its activities. Then, this culture defines their presence in the public eye—that is, how the people perceive their station.

Altering fire service culture is more complex than changing the type of hose carried by a department's fire engine. Changing a group's culture—the combination of rituals, values, and traditions that define them—is challenging. The existing culture endures, resisting change, because it has been a critical component of the glue that holds the organization together.

Organizations and teams that are successful with change management and culture shifts quickly recognize and respond to signs of change. In other words, they show *organizational resilience*. They've worked out how to experiment fast, often, and cheaply, not only with products and services but also with fire department structures, methods, and tactics. They've polished their skills in managing complex multiple-stakeholder systems in an increasingly interconnected world. Perhaps crucially, they've discovered how to capitalize on their most precious asset: their staff.

Firefighters have come a long way in the past couple of decades. Although it's clear that much needs to be done to ensure that the fire service is on par with other emergency services, it cannot be denied that there has been transformation in the fire culture that has allowed it to progress. Next let's explore the key areas of development where shifts in fire service culture are evident.

Increased Focus on Firefighter Health

Firefighters are at increased risk of cardiac-related deaths, among other challenges to their long-term health and avoidance of injuries. Continuous exposure to combustion products and other chemicals—including toxic substances such as polycyclic aromatic hydrocarbons—that might be present at firefighting locations also increases their risk of cancers, including leukemia and other types of cancer. Many organizations and industrial leaders are studying this to reduce deaths and ensure that firefighters get a fair chance of living their lives after serving their communities.

Prioritizing firefighter health—and studying ways to maintain and improve it—is essential so that members can continue their service to the community and, ultimately, their lives beyond their service. Addressing the increased risks and challenges firefighters face because of their work should become part of the culture. Fortunately, this is already being seen with comprehensive initiatives to ensure that agencies and organizations begin properly tackling this critical challenge.

To this end, the National Fire Protection Association (NFPA) implemented their *Standard on Comprehensive Occupational Medical Program for Fire Departments* (NFPA 1582) in 2020, with an update in 2022. NFPA 1582 calls for a comprehensive occupational medical program to boost firefighters' health and ensure that they are healthy enough to work. For example, it encourages firefighters to make lifestyle changes to reduce the risk of heart attacks. Such initiatives become integrated into the organizational culture to help everyone by prioritizing a healthier work environment; this benefits the efficiency and health of the entire agency, as will be explored later (see "A Deeper Look into the Culture of Fire and Emergency Services").

Community Risk Reduction Programs

Another critical shift has been the implementation of community risk reduction (CRR) programs. A CRR program was first outlined in NFPA 1300 (*Standard on Community Risk Assessment and Community Risk Reduction Plan Development*) in 2020. It outlines a process in which communities can outline fire-related risks and then plan to eradicate the threat. This process includes risk assessment, developing a proper risk reduction (i.e., CRR) plan, and then implementing and evaluating that plan. All in all, NFPA 1300 provides guidance and prioritizes local risk reduction and the allocation of resources to ensure that communities are adequately prepared. Fire departments all over the United States have implemented CRR programs to create awareness related to fire incidents in homes, offices, and schools.[32]

The CRR model has turned out to be very successful. For example, one CRR model that was used to create a fire prevention education program reduced the number of fire calls by 35%.[33] The more fire departments implement this, the more lives will be saved, and there will be less burden on the existing firefighters, future firefighters, and the entire community. By implementing further CRR programs as part of organizational culture transformation, firefighters have seen beneficial shifts in their efficiency and impact as an organization.

More Accountability

Today's firefighters have also embraced further accountability on the job. There was a time when firefighters weren't held accountable. However, that's all changing. According to the *2018 ESO Fire Trends Report*, fire departments will face even more scrutiny regarding the allocated budget.[34]

Since communities are restricting their budgets and fire departments are being held accountable for their services, they will have to provide evidence to continue their funding. Otherwise, they will face more cuts. This will—and already does—significantly influence organizational culture.

In many fire service organizations, fire departments' and team members' contributions are assessed through various accountability measures. These establish structures and lines of support to ensure that firefighters have clear lines of communication, understand their roles, and commit to their duties.[35]

Accountability procedures range from personnel accountability systems to incident command systems. Specific measures that have been implemented by some organizations include check-in and checkout procedures to ensure that there is an accurate count of personnel at all times.[36] After-action reviews also play a crucial role in measuring the team's effectiveness and overall response to the incident.

These measures have changed the way that not only independent fire stations but also individuals are held accountable.[37] Coupled with further oversight of their budgetary allocations and uses, fire departments are accountable to the community for each incident they respond to, making them even more efficient. This is paired with an increased emphasis on training, as will be discussed next. Hence, the fire service culture has shifted drastically—and for the better.

Executive Training

The fire chief's role has also changed over the past few decades. There was a time when the only role a fire chief had was being a leader during a fire response. That's not the case anymore. Fire departments must take on additional roles to stay afloat. They are expected to respond to both human-made and natural disasters. From responding to terrorist attacks to large-scale incidents such as transportation accidents, fire departments must train to fight fires and protect their communities from other threats.

As a result, if they want to advance in the ranks, firefighters are now expected to complete executive training to show they have plenty of leadership skills to navigate an emergency. Available training programs include the International Association of Fire Chiefs' Officer Development Program, which focuses on the development of leadership, as well as management skills for various levels of the department. With a standardized curriculum, officers become trained for every stage of their careers.[38]

Additionally, there are several other programs that firefighters and officers can engage in to ensure their progression within the field. Powered by the Federal Emergency Management Agency (FEMA), many in the industry already know of the National Fire Academy, with courses designed to assist officers in preparing themselves for all situations and challenges they might face on the job. Given its association with FEMA, the National Fire Academy curriculum includes details, situations, and procedures that expand upon the usual expected situations a firefighter might encounter. This ensures that each officer is well rounded and prepared for all challenges that firefighters respond to.[39]

With such significant changes, these initiatives have changed the nature of organizational culture and expanded the experience and knowledge necessary to be a firefighter. Nonetheless, as we will see next, challenges and consequences come with changes to fire service culture.[40]

The Consequences of the Change in Culture

The change in the culture is both good and bad for the firefighters. For starters, implementing a proper health care program for the officers will help firefighters not only on the job but also off the job, when they are out living their lives with their loved ones. If they are in better shape, they can give more back to the community and worry less about being alive and fit enough to watch their children graduate from high school.

Another pro is the implementation of community programs. If more people are aware of the risks associated with fires, there will be fewer fires and less damage to property. Moreover, fewer people will succumb to the fire. For the firefighters and the public, this is of great benefit. It will not only increase the community's preparedness but also ensure that even staff with limited support and shortages can handle each incident they respond to.

There are cons with the change in culture as well, though. Since there is more accountability, firefighters are now expected to give concrete evidence of their actions and how they save lives. If they can't produce it, their departments will face budget cuts. This will make it even more difficult for existing fire stations to recruit newer people or train them properly because there is no concept of job security. If the budget cuts continue, people will be forced to become part-time firefighters or volunteers. With the current challenges of finding and retaining volunteers, it is unlikely that relying solely on volunteer firefighters is the best plan of action to prepare and support a community with the proper emergency services.

While switching to a part-time or volunteer department might sound like an ideal solution, by reducing the funding necessary to support the usual staff in these departments, this is far from a positive solution when all potential ramifications are considered. When there is a fire emergency, fire chiefs require a lot of people, not just the limited staff they have on hand. Not only is it difficult to rely on part-time firefighters and volunteers, but it also increases the workload on existing firefighters who are working round the clock—and maybe off the clock.

As mentioned, several departments are currently unable to respond to incidents during daytime hours. If these fire stations relied only on volunteer firefighters, the reliability would worsen, making it clear that this is not what the future should hold. This leaves one other important area to discuss: the role of leaders in facilitating these significant cultural and organizational changes, now and into the future.

The Role of a Leader

A lot needs to be done to change the fire service culture—and to ensure that this new culture is maintained in the years to come, to ensure positive results for future firefighters. A leader must step up their game to facilitate the change—a feat that isn't easy. Without the leader's guidance and facilitation of this effort, it is unlikely that significant change will happen.

In Frank Viscuso's seminal work *Step Up and Lead*, he outlines several principles of effective leadership, particularly within the context of emergency services but applicable across various fields. One of Viscuso's central principles is the importance of leading by example.[41] This principle underscores the idea that leaders must embody the values, behaviors, and work ethic that they expect from their team members.

A practical and attentive leader must maintain situational awareness wherever their personnel are being led. Additionally, a leader must overcome the habit of being selfish in the workplace; a benefit to any one team member would benefit the entire workforce and organization. Instead of allowing selfishness to rule, leaders must share their vision with others and exhibit it in front of them; this is also known as *vision casting.*

Dr. Edgar Schein's iceberg analogy (fig. 2–1) can be used to explain fire service culture and how much fire chiefs must understand about their departments. Picture an iceberg submerged in water. Only the tip of the iceberg is visible, while the rest—a substantial proportion—is submerged in the water. The portion that is above the water represents the components of the fire service culture that anyone can see—namely the uniform, expressions, hierarchy, and the general equipment of the fire station.[42]

The part of the iceberg submerged in water is where the battle lies; that's the organizational culture. It's implicit and includes codified processes that are built over several generations. If you go deeper, that's where the crux of the organization is. That's where all the values, the beliefs, and the assumptions as part of workplace culture exist, and it's often overlooked. That's where a firefighter can tell you if they are refuted or supported by the culture. These features define who firefighters are at their core—and that's not something a civilian can see.

Future fire chiefs must understand the concept of culture and what it represents to succeed at making a change. If they misunderstand the idea, it will only create more chaos. But here are the critical questions for fire chiefs: If they understand the concept, how can they bring about change? More important, if they have brought the change, how can they ensure that it will last for a long time?

Figure 2–1. Dr. Edgar Schein's iceberg model of culture is a framework that explains organizational culture by using the metaphor of an iceberg. This model suggests that much of an organization's culture is hidden below the surface, just as the bulk of an iceberg lies beneath the water.

Business Continuity Management

One theory about cultural development and maintenance is *business continuity management* (BCM). This is a holistic management process helping organizations to identify threats and develop strategies to ensure that they continue to deliver critical department functions before—and after—disruptive events. For the fire service, BCM has incredible relevance, and this concept determines how fire stations embrace organizational cultural change.[43]

Many activities related to BCM are critical for any organization, including a fire department, to overcome disruptions and alterations to their organization. By utilizing risk assessments, impact analyses, business continuity planning, crisis management, and a lot of trial and error through continuous improvement, organizations overcome the disruptions that would otherwise cause challenges for them in the future. For disaster recovery planning, BCM is integral to the success of the entire organization, as well as their ability to support the community at large.[44] BCM plays a vital role in developing and completing cultural reform for the fire service.

Behaviorism

Aside from BCM, three pillars allow change to occur: behaviorism, constructivism, and methodologies. If you align all three pillars, you can go a long way in implementing and maintaining change.

Have you ever heard of the saying that to change the culture, you must change the behavior? It's simple enough to understand and implement. *Behaviorism* is the theory that all behaviors are learned through conditioning. Pavlov's famous experiment demonstrated *classical conditioning*: specific cues produced specific responses. Skinner took this a step further with *operant conditioning*, which showed how such cues could reinforce consequences, including rewards and punishments.[45] Therefore, in other words, we see or experience, and then we learn.

Behaviorism is something we see in small children as well. If children see their parents fighting constantly, they'll grow up thinking it's the only way to solve conflicts. However, if a child is taught that problems can be solved rationally without resorting to screaming and throwing things, that's what they will learn. What we are shown and allowed to repeat then becomes a behavior that we cultivate, for better or for worse.

To bring change, use behaviorism as a tool. You can study the culture and see what behavior is approved and what is reproached. However, you must be attentive to the culture and study the behavior of your subordinates closely if you want the desired results. Once you have identified them, you can formulate a strategy to change them or ensure other people follow it. Use rewards and other responses to ensure that your staff knows what is right from what is wrong or to adopt a new cultural standard within your organization. This will make it easier for you to align with a fire station's organizational values and help ensure that your staff has a chance to adapt and understand the new expectations within the workplace.

Constructivism

Constructivism states that people form their own opinions about the organization based on what they have understood and experienced while working there. Constructivism is another pillar to master if you want to bring lasting change. Cultural change can be implemented through constructivism when people learn and understand the values of the organization.

Constructivism is a promising paradigm for learning and teaching. Since it follows a "see one, do one, teach one" approach, it is more effective at implementing change. Constructivism encourages learners to explore and take part in an active process. Therefore, it can be much more effective than other paradigms. While prior knowledge can influence their learning process, learners are still encouraged to make connections and collaborate with others, which is paramount in a team setting like the fire service.[46]

Fire chiefs can use constructivism to forge a new and more straightforward path to the future. However, for the plan to be successful, you must iron out the details and make sure everyone knows why there is a need to change the current culture and how to go about it. You must ensure that people understand that this is a positive change. If you don't plan the whole shift properly, people on your team who resent the idea of change might make change more difficult for you to implement.

Methodologies

When implementing a change in the culture, you also must be sure you have a transparent system of tools to make the change more successful. Fortunately, there are plenty of options to choose from: You have a whole set of *methodologies* at your disposal. Methodologies entail the *what* and *how* of the organization. From emergency responses to meetings to everyday work, a methodology will include just about everything.

Before implementing change, you must see how the existing methodologies work. You will also have to know what drives outcomes and decisions in your workplace. Moreover, you will need to see if your organization has a collaborative decision-making process or if significant decisions are taken at the top. All of this is crucial to understanding the methodology and will help you in the long run.

As a leader, you must keep your head in the game and ensure that the culture manifests appropriately. It's not going to be easy. Simply put, change isn't easy; it will take a lot of time and effort to get there. However, it's also crucial to remember that several obstacles relating to change management can be eliminated or alleviated by being honest and transparent with the rest of the team.

Unfortunately, that's where people may flake out. They don't trust the team enough to let them in on what's going on. Or even if they do, they believe it's best to keep planned changes hidden until the time comes to implement them, to prevent backlash. For a leader who believes in the autocratic approach, this may be less of a problem. According to them, the subordinates must do something simply because they must do it. There is no way around it, in their view.

By contrast, it's your responsibility as the (nonautocratic) leader to reach out to people and explain what is happening. This approach will mitigate anxiety and ensure that the change goes smoothly. The key is to engage with your subordinates early instead of springing it on them at the last second.

On the one hand, some people will accept the change and be optimistic about it; these are the ones who will give you no trouble. On the other hand, you will have people who reject the change entirely. They are the ones you should watch out for and prepare a strategy to deal with them. Remember that such feeling often comes from a place of loss and fear; some people just aren't comfortable leaving their comfort zones and moving on. This is not a bad thing either—in fact, it's arguably natural. Still, it's up to you, as the leader, to navigate the situation and get to a place where the organization can transition smoothly.

How to Maintain a Healthy Organization

Once you have implemented the change, another problem arises: maintaining a healthy organization. As I mentioned earlier, not many people are open to the idea of change. You might have overcome resistance, but you probably haven't. It is evident when you encounter resistance: Some subordinates attempt to adapt but want to go back to the old culture and do away with the new one entirely; others straight up reject what you have presented, stubbornly holding onto the past tenets of your old organizational culture. Thus, there will be a spectrum of responses across the team; truthfully, you never quite know how your team will respond until they do.

You must proactively address this resistance by aligning with community members. These include elected officials, management, volunteers, and other stakeholders. You can set up a process through which you can solicit regular feedback and address issues. You can also sort out any concerns and ensure that no one feels left out. This will help to create a sense of adaptability, innovation, and inclusion. As Heath asserts, this is a perfect upstream solution to start with, to prevent challenges that might arise later. As he explains, not every upstream solution will be correct, but it will always be intentional.[47] Every step of the way, you need to include the stakeholders who will be affected. By including them, you avoid misunderstandings and ensure smoothness in implementing your changes for the community and your fire department.

You can also promote people who show the desired organizational behavior. This will act as an incentive for people to resist the idea altogether. Much like our discussion on conditioning, you can present rewards for those assisting you in progressing into a new era for your organizational culture. You can also recruit people who align with the new culture and have the abilities and skill sets to promote the change. You can crowd out the challenges you are facing in several ways but usher in the new positive atmosphere and changes you desire for your entire staff.

Moreover, you can focus on education of employees and ensure that they develop professionally. This means investing in their training and promoting them to leadership positions. This is one of the best ways to ensure that your staff is on the same page with you and to encourage them as a team to embrace the changes to achieve and maintain the new organizational culture. Providing education and chances for employees to develop might further motivate even the most reluctant of your team members. Furthermore, you can be a mentor—or even implement a *reverse mentorship* program where senior leaders mentor the recruits in the job, while the recruits help senior members with the latest technology. These are excellent ideas that include your team members—and just as important ensure that they feel included and that you remain connected with the people most impacted by the changes you make as a leader.

Above all, don't forget to establish a process where you can constantly assess individuals' knowledge, skills, and abilities and use them to mold your methodology. Most importantly, your process should match and mirror the way that your team learns, adapts, and grows. When implementing change, ensure that your methodologies reach your team members efficiently and without confusion.

Be patient and consistent when implementing change. Successful transformations cannot occur if your team members are not on board. Thus, devote time to mentor them, be honest with them about your apprehensions, and keep them in the loop. If you don't, the whole initiative might fall flat, and you would have missed your chance to bring about monumental, transformative change.

Chapter 2 Review Questions

1. Which of the following is NOT a way to prevent resistance to change in the organizational culture?
   a. Crowd out challenges and usher in a new positive atmosphere
   b. Align with community members and get regular feedback
   c. Promote people who resist the idea altogether
   d. Be a mentor or implement a reverse mentorship program

2. How does mission-driven culture make the ability for feedback and employee development more efficient?
    a. By avoiding measuring operational outcomes
    b. By not recognizing performance appropriately
    c. By not focusing on the mission
    d. By appropriately measuring operational outcomes and recognizing performance

3. Which of the following describes how to keep team members in the loop during the change process?
    a. Excluding them from discussions
    b. Rushing through the change process
    c. Promoting a rigid hierarchical structure
    d. Mentoring them, discussing apprehensions, and maintaining communication

4. How does mission-driven culture help leaders focus on their priorities within the organization?
    a. By increasing micromanagement
    b. By allowing leaders to focus on frivolous aspects
    c. By ensuring strategic alignment and balancing priorities
    d. By promoting decision-making on every aspect

5. How does constructivism differ from behaviorism in promoting change in organizations?
    a. Constructivism focuses on rewards, while behaviorism focuses on cultural values.
    b. Behaviorism follows a "see one, do one, teach one" approach, while constructivism does not.
    c. Constructivism encourages learners to be active participants in the learning process.
    d. Behaviorism focuses on what people understand and experience, while constructivism does not.

6. Cultural change can be implemented effectively through constructivism by ensuring that people understand the values of the organization.
    a. True
    b. False

7. Mission-driven cultures in fire departments can lead to an increase in job satisfaction for firefighters.
   a. True
   b. False

8. A leader who believes in the autocratic approach may face resistance from subordinates during change.
   a. True
   b. False

9. Constructivism encourages learners to explore and take part in an active learning process.
   a. True
   b. False

10. A key aspect of mission-driven culture is that it is driven by individual goals and the overall mission of the organization.
    a. True
    b. False

11. How important is the role of leaders in shaping the culture and principles of a fire department?

12. How does the fire and emergency services culture promote continuous learning and growth among its members?

13. What are the key factors to consider when formulating a strategy for implementing change within an organization?

14. How does the fire and emergency services culture inspire new recruits to embody the values of teamwork and service to others?

15. How does mission-driven culture impact the organizational structure of fire departments?

3 DILEMMAS, DISSONANCE, ETHICS, AND LEGAL CONCERNS IN THE FIRE SERVICE

Ethics in 21st-Century Organizations

Ethics are a top focus for any organization. Ethics are neither a luxury nor an option in the 21st century. In a world where environmental and diversity concerns have become integrated into the fabric of organizations, applying ethics is not as much of a question as planning how to use every facet required for your team. There is growing dissatisfaction with selfish and irresponsible behavior within society that impoverishes some while enriching the cunning. In modern society, the acknowledgment of ethics as a priority benefits various industries including, importantly, the fire service, and it influences how departments operate.

Under Carly Fiorina's leadership as chief executive officer (CEO), Hewlett-Packard established its commitment to ethical integrity. "Some of the most critical decisions I ever made were terminating people who weren't conducting themselves with integrity," the former CEO of Hewlett-Packard remarked of her 6-year tenure.[1] R. J. Reynolds Tobacco, by contrast, was found to have concealed information about nicotine's addictiveness. Even though this was done within the industry's legality, it is challenging to justify this practice ethically, given its drastic impact on society. In the 21st century, this sort of conduct is no longer acceptable. Tolerating bad behavior to improve the bottom line and explaining it as "not actually illegal" creates a toxic environment, regardless of which industry you are part of.

There are other compelling reasons for 21st-century enterprises to prioritize ethical ideals and social responsibility. This is the topic of this chapter.

It's All About Responsibility

Understanding organizational ethics necessitates a discussion of responsibility. It may seem self-evident, but irresponsible behavior is prevalent and must be recognized and addressed. Greed can force leaders to lose perspective and rationalize unethical behavior fueled by personal ambition. Examples are CEOs who drive their employees to reach sales metrics while failing to adhere to organizational standards and bylaws for operating and marketing to their customers. These practices are not isolated to any one industry and unfortunately are frequent in some companies. The bad news for criminals is that society is more aware than ever of ethical irresponsibility—that is, what helps some while (or by) harming others.

According to Kilcullen and Kooistra, *business ethics* can be defined as "a system of principles that directs company actions to show a concern for society as a whole while pursuing profits."[2] The phrase "concern for society as a whole" encapsulates the current, growing dissatisfaction with many selfish, irresponsible behaviors by organizations.

Concern for society as a whole is best recognized as the hyperfocus on the need for corporate social responsibility. This refers to organizations' ethical and responsible actions and ensuring that they contribute positively to their society, community, and the environment. Organizations across industries have been engaging in this conscientious practice for years, many of them creating divisions dedicated to sustainability practices while others are focused more on charitable donations than to their stakeholders.[3]

For the fire service, corporate social responsibility also plays a crucial role in how each organization should operate and the kind of initiatives that leadership can promote. Fire departments—many of them, to my knowledge—continue to work using this very principle today. They engage in community education, outreach, and social and environmental responsibility. Fire departments show their communities important fire safety tips, participate in community events, and even partner with nonprofits. Many departments also engage in energy-efficient practices already, reducing their environmental impact.

The fire service goes beyond these initial outreach programs and promotes diversity, equity, and inclusion (DEI). Efforts to include all individuals in the fire service and to become an inclusive force are a present reality. As with any industry, DEI initiatives are not optional for the fire service; rather, they have become a responsibility that departments commit to because they are concerned for society and motivated to serve their communities. This movement is about commitment to doing what is right, as well as ethics.

Ethics Are a Priority

Ethics must be an organizational priority, rather than just a legal or moral obligation. In today's highly competitive environment, corporate leaders have a lot to think about. To name a few examples, they must keep up with rapid technology breakthroughs, competitors' products and services, globalization's consequences, and possibilities and risks within their industry. Leaders must also continually monitor their organizations' purpose, vision, values, culture, strategy, and goals.

Given all this complexity, finding room on the organizational plate for another key priority may be challenging. However, to succeed in the 21st century, organizations must figure out how to prioritize ethics. The challenges of DEI, among other essential values, tie into the importance of ethics toward the organizational culture.

Priorities are the items that are judged to be the most significant. For fire service professionals, their priorities typically are related to safety, teamwork, professionalism, accountability, and continuous improvement. At times, all other values are secondary to these; however, ethics must be among the primary functions within the organizational culture. Ethical values must become recognized as one of the most significant success factors for the modern fire service.

Ethics will influence decision-making and institutional culture as an organizational priority. An alignment approach that connects business ethics with purpose, vision, values, strategies, and goals is required to accomplish this ideal. Because ethical ideals are fundamentally social, this alignment approach will focus on connections and creating relational expectations. The improvement and support of all is the purpose of an ethical organizational culture. External ties with clients, consumers, vendors, and the community and internal relationships between leaders and followers are all valued. As a result, people are regularly treated well, and an ethical culture develops.

Beyond Responsibility and Prioritizing Opportunity

Today's fire service organizations must go beyond ethics to protect their brand, avoid negative press, and comply with mandated rules. Organizations that recognize the importance of ethical ideals in influencing the future will have a considerable advantage.

Every organization can affect social culture by institutionalizing ethical ideals, given society's critical need for one. When this happens, communities benefit from the good influences that employees bring home to their families, friends, and colleagues.

Many fire departments recognize the importance of ethics in their organizational culture. There is, however, still a lot of opportunity for development. First and foremost, the fire service must regard ethics as a social duty to which they are bound. Harming others for personal gain, no matter how skillfully done, is unacceptable. Once this is accepted, and only then, will ethics find its proper place in organizations by establishing relational norms that benefit everyone. Finally, the most successful fire agencies will seize the opportunity to mold the future by instilling ethical ideals in the culture. Building a values-based culture can improve positive influences that propagate outside the organization.

Why Are Old Models of Ethics Outdated?

Multinational organizations recognize that a robust code of conduct or code of ethics is an essential compliance instrument in today's international marketplace. However, an out-of-date legacy code can pose a serious compliance risk. Even so, updating codes of ethics can be difficult, often being time consuming or complex beyond measure.

Creating or just amending a worldwide code of ethics to bring it up to date is a surprisingly nuanced task. Rushing in with a careless, just get it done strategy, such as renaming a U.S. employee handbook as a global code or copying and pasting from other organizations' conduct codes, risks infecting the organization with global compliance flaws. With the intricacies of the modern principles of ethics, it is essential to provide in-depth review, revision, and updating to ensure that the content matches expectations and does not present any missing elements.

The obstacles and contradictions involved in imposing a single set of standards for employees across numerous jurisdictions necessitate a well-thought-out plan for creating an effective global conduct code. Multijurisdictional organizations should have a strategy that takes into consideration common issues. Yet, a constant challenge exists when attempting to develop universalism with your principles of ethics while also trying to tailor each fire agency's ethics to their location and regional expectations.

Expect a tug-of-war between global uniformity and the parochial norms and local laws of each individual jurisdiction when you export a code of conduct or ethics across regional or national operations. You may need to issue aligned local regulations or a local code rider for each state. Multinational organizations, however, consider their ethics codes as intrinsically worldwide in scope; thus, their headquarters tend to reject country-by-country modifications because they seek a single code to remember that champions the organization's essential ethical norms. In many ways, the initial efforts and difficulty they might find in creating their universal ethics codes do pay off, as their regulations are applicable to several nations and regions. Yet, there is always the potential challenge that ethics codes of different states—or even within the same county—will be incongruent and incompatible, posing another critical challenge to organizations that attempt to impose a multinational ethics code.

This consistent strategy necessitates difficult trade-offs. Specific rules in an ethics code may be legal within the jurisdiction of the agency, but they may be illegal in other areas of the country. For example, a mandatory reporting regulation forcing employees to report coworkers who break corporate policy may be unenforceable in areas outside the jurisdiction of the agency. Furthermore, rules in a code may comply with the legislation in the jurisdiction where the department is located. Still, they may be impossible to harmonize across several states (think of the listing of protected groups in a global antidiscrimination provision). Furthermore, a code's regulations may be illegal under state law. For example, a diversity requirement requiring workforce data tracking could be disallowed in one state but acceptable in another.

Ultimately, many ethics codes are outdated owing to time or locale. With this realization, the solution becomes apparent to overcome ethical dilemmas and modern ethics challenges—namely to update or replace these with modern ethics codes.

Overcoming Dilemmas and Prioritizing Modern Ethics Codes

Modern ethics codes are tailored to applicable standards that fire departments should be held to. However, several dilemmas could arise in the creation of a modern ethics code. These warrant consideration since it is only through the thorough examination of the ethics and the priorities of the fire service agency that significant change can occur.

One dilemma is that the fire service struggles with balancing the moral standard of serving the community while ensuring individual safety. Yet, a fire department code of ethics and moral education on safety and health can prioritize both appropriately. By adhering to local expectations, studying other local ethics codes, and even considering the organization's priorities, fire departments can pave a pathway to prioritize modern ethical standards.

Another critical challenge is ensuring that DEI issues are overcome through more inclusive policies. That is already happening around the country and the world: female and male quarters are being created, and more inclusive policies are being formalized in their stations. Change is coming, and there's plenty that you and your organization can do to become part of forward progress in the future.

Ethical dilemmas in the fire service require capitalizing on opportunities and modernizing the standards that all members are held to. This means regular training, ethics committees, and transparent communication with the community. Furthermore, all stakeholders must be included when prioritizing organizational change. With a balance of priorities and constant mitigation of these challenges, a fire chief can lead their department to success and ensure that they do so within the guidelines of what is considered ethically standard today.

Understanding a Fire Department's Legal Concerns and Challenges

Once the ethic codes are dealt with and fire chiefs have created plans of action to overcome the various dilemmas that their organization has faced and will face, another crucial step that the organization must take to pave a positive pathway forward—for both them and their stakeholders—is to learn how to manage legal challenges that they might encounter.

The fire department is the most common resource for local-level disaster management worldwide. This should come as no surprise, given that fire is the most common daily hazard in almost every society globally. Early fire departments served a single function, but as time passed, many countries' fire departments developed their skills to tackle a wider variety of both common and rare threats, such as the following:

- Extinguishing a fire (structural fire, brush fire, wildfire, and hazardous materials fire)
- Disaster response coordination
- Fire and arson investigation
- Extraction of a vehicle
- Reanimation (urban, swift water, wilderness, cave, airborne, alpine, dive, crack and crevice)
- Reaction to terrorism (actual or threatened)
- Hazardous materials response and cleanup
- Warning issuance
- Emergency medical services (EMS)
- Emergency management
- Fire and structural safety inspections
- Prevention
- Community risk reduction (CRR)
- Training

Local, regional (e.g., county or province), and national fire departments exist. Whether firefighters are paid versus volunteers and the kind of government in existence usually determine the organization of a fire department. The three types of fire department organizational levels (with examples of countries that employ each type) are as follows:

- On-the-ground organizational structure (Canada, Germany, United States)
- Region's organizational structure (Australia, United Kingdome)
- Country's administrative structure (Spain, France, Hong Kong, Israel)

Both community threat and prior funding availability determine fire agencies' needs and response capabilities. Domestic cooking and heating techniques, outdoor burning habits, home materials, design, population density, industrial operations, climate, vegetation, water availability, and geography all influence the likelihood of a fire. Funding sources and levels differ by country, but they are seldom adequate to manage known fire threats and meet all the fire department's objectives.

For example, Article XI, Section 11, of the Washington State Constitution authorizes counties, towns, and municipalities to enact "local police, sanitary, and other regulations," but no specific provision mandates them to provide fire protection services. Furthermore, a port district is not required to provide fire protection services. In practice, however, most regions offer fire protection in one of the following ways:[4]

- Maintaining an entirely agency-staffed fire department
- Forming a fire protection district (counties, cities, and municipalities can all form a district)
- Contracting for fire protection services with another city or fire protection district
- Annexing immediately to a fire protection district with voter approval
- Obtaining voter support for the establishment of a regional fire protection authority

Yet, these legal concerns are only the tip of the iceberg. One significant barrier to fire departments' actualizing the scope of their plans remains: To modernize—and to continue to protect citizens—funding is required, which is limited in availability today.

Limited Fire Department Funding: Where Does It Go?

Every day, firefighters face considerable challenges while trying to keep their communities safe. Fire chiefs must ensure that their employees are always ready to react while minimizing response times.

Fire departments and EMS are critical to U.S. communities. Every year, they respond to millions of calls for assistance. As natural disasters, terrorist attacks, and hazardous material spills become more regular, government funding is becoming increasingly scarce. This implies that firefighters will be given more responsibility but fewer resources to assist them. Today, more than ever, fire departments must seek money from sources other than state and local governments, which are traditionally held accountable for fire services.

For many fire departments, much of their funding comes from organizations and federal agencies, including FEMA. There are numerous FEMA grants that fire departments can apply for; however, the enrollment window is short, and funds fluctuate from year to year, often at the discretion of the congressional bodies that oversee the federal budget. The Assistance to Firefighters Grants program supports firefighters all over the United States and can provide essential assistance when available. The federal government has issued billions of dollars in grants to local stations, providing additional funding for their efforts.[5]

Aside from federal agencies, there are state and local funding sources. Many of these are derived from general tax revenues delivered to public safety and essential services departments, including the fire service. Property taxes, sales taxes, and more are all collected and then allocated for use by these departments. If this is not enough—especially given regular cuts and alterations to state and federal budgets—fire departments do have alternatives.[6]

Fire departments can raise money through community fundraising and other initiatives. Many departments see success with these efforts because of their direct access to the community, and such funding can bridge the gap between citizens and the fire service. Volunteer fire assistance programs also lessen the burden that operational costs impose on many fire stations; however, even these funds may simply not be enough, and shortfalls have a drastic impact on the efficiency and support of the organization.[7]

Budget cuts pose a direct challenge for fire stations to serve their communities, and these have left many fire departments smaller towns understaffed since—and even before—the COVID-19 pandemic. Some fire service agencies have been forced to downsize their organization or shut down completely. Funding challenges leave fire service administrators battling to maintain adequate training, policy compliance, equipment upkeep, and most critically, the safety of responding firefighters and their communities.

The prospect of transitioning from traditional communication technologies to mobile applications and cloud-based platforms exists. Firefighters rely on quick, accurate information to make split-second decisions, and their commanders are routinely flooded with data that might be difficult to go through. By their function as a force multiplier, mobile platforms can assist fire chiefs and personnel in progress against these other operational issues.

According to the U.S. Fire Administration, there are around 1.2 million firefighters in the United States. About 900,000 of them are volunteer or volunteer-and-work firefighters. Considering this, fire chiefs of volunteer organizations often rely on ten or fewer firefighters on high-risk events. As a result of the low staffing, all residents and firefighters are at risk. Many communities continue to slash funding for professional fire services, resulting in higher call levels with fewer personnel.

The Occupational Safety and Health Administration (OSHA), within the U.S. Department of Labor, requires that when two firefighters fight a fire inside a structure, two others must be standing outside, ready to rescue them if needed. On-scene firefighting, search and rescue, ventilation, water supply, and speedy response are frequently insufficient. The OSHA guidelines, as beneficial as they are for ensuring safety, are challenging to follow when there are limited volunteers and even limited staff. This is also coupled with the intense training required by the OSHA mandate, much of which can be intimidating for those who want to join the fire service. Limited staffing is further exacerbated, and the overall impact on public safety is the most significant detriment for citizens and firefighters.

A busy schedule and family responsibilities leave few volunteer firefighters with time to complete the rigorous training, which is why retention is a major concern in volunteer and combination departments—not just the retention of volunteers but also the retention of paid firefighters. According to FEMA, North Carolina's volunteer firefighter population has fallen by 22% in 2 years. Similar shortages exist in other states.[8]

Even the U.S. Forest Service is struggling with shortages of staff. Their inability to recruit—partly because of the pandemic, as well as the dangerous nature of the job—has left them in a critical place, without many resources amid intense wildfires. In many cases, this national firefighting department must reach out to already understaffed local fire departments for support, but they struggle nonetheless.[9]

Legal Challenges Facing the Fire Service

Across decades in the fire service, several important legal considerations have been constant themes. Aside from the issue of funding, fire service leaders will face continuous challenges keeping up with health and safety regulations, such as the OSHA mandate in the United States. Arguably, funding issues only exacerbate preexisting problems including noncompliance with OSHA regulations and other regulatory or industry standards.

One critical legal challenge in the fire service is the potential for civil rights violations. As will be explored in chapter 11, by employing proper DEI practices, fire departments can prevent civil rights lawsuits stemming from discrimination, unequal employment opportunities, and harassment—all serious allegations that have been issues within the fire service. With appropriate values, leadership, and organizational culture, these issues can be avoided, a goal every fire service leader must aspire to.

Unfortunately, lawsuits are always possible following the service that the fire station performs. Firefighters commit themselves to serving the community, but at times, firefighters can be held responsible for the most troubling results of fires—rightfully (in the worst of situations) or wrongfully. Property damage, negligence, and wrongful death lawsuits are all legal issues that can be raised because of what happens in the field. However, steps can be taken to avoid lawsuits and legal malfeasance.

First, as a fire service leader, negligence and any concern for legal accountability can be addressed proactively by following standard operating procedures. Furthermore, property damage can also be mitigated with a proper understanding of required fire suppression efforts and the correct protocols. Not every legal issue is avoidable, but you can prevent or solve most legal concerns through proper leadership, training, and education.

The Importance of Preparation

With proper understanding and oversight—and especially with good leadership—most legal concerns can be easily prevented. While funding concerns remain a consistent challenge to overcome, the future of the fire service seems promising in terms of essential changes that can be made. Legal concerns will be minimized with more oversight, proper accountability, and education, paving the way for further success. As Heath explains, by applying the tenets of upstream thinking, organizations in the fire service can prepare and prevent problems and excel in the line of duty.[10] In the face of legal challenges, education is a crucial component to preparing your team, as well as critical to brand management, as we will discuss in the next chapter.

Chapter 3 Review Questions

1. Ethics codes become incongruent and incompatible across different states or regions because of which of the following?
    a. Lack of training for employees
    b. Failure to enforce policies uniformly
    c. Varying legal requirements
    d. Inconsistencies in employee benefits
    e. Language barriers

2. Which of the following is a critical legal challenge that can happen following the service performed by fire stations in the community?
   a. Unfair employment opportunities
   b. Equal employment practices
   c. Diversity and inclusion policies
   d. Incorrect promotion of standard procedures
   e. Negligence, property damage, and wrongful death lawsuits

3. Why must ethics become an organizational priority in the 21st century?
   a. To avoid legal consequences
   b. To increase profits
   c. To emphasize diversity and inclusion
   d. To succeed in a competitive environment

4. Which of the following is a significant concern in volunteer and combination fire departments regarding staffing?
   a. Overstaffing
   b. Intensive training requirements
   c. Rigorous recruitment process
   d. Low retention rates
   e. Excessive funding

5. What kinds of factors influence the likelihood of a fire?
   a. Population density
   b. Industrial operations
   c. Climate
   d. All of the above

6. Legal concerns can be minimized with oversight, accountability, and education in the fire service.
   a. True
   b. False

7. Heath suggests that applying upstream thinking in organizations can help prevent legal issues in the fire service.
   a. True
   b. False

8. Property damage, negligence, and wrongful death lawsuits are not common legal issues in the fire service.
   a. True
   b. False

9. Preparing a fire department with proper leadership can prevent most legal concerns.
   a. True
   b. False

10. Communication of expectations and reasons by leadership is not important in fire suppression efforts.
    a. True
    b. False

11. What challenges do fire service agencies face in maintaining adequate staffing levels, especially in smaller towns?

12. How do funding sources and levels for fire departments vary, and what challenges do they face in managing fire threats?

13. In what ways can oversight and accountability help in addressing legal issues within fire departments?

14. Why should ethics be considered a priority rather than just a legal or moral obligation for organizations?

15. What legal issues might firefighters face in the field?

4 MANAGING THE BRAND

The main goal of brand management is to maintain a good perception of your organization among the target audience. With a positive brand image, your organization is likely to be received warmly in the community. Every part of your marketing and content strategy must align with your *brand*—that is, the offerings and image that you want your community to see of your fire department. Yet, establishing a brand is not a one-time, quick fix. This entails a long-term plan, and many often underestimate the scope required for managing the brand.

A brand does not emerge overnight; it takes time, patience, and the employment of a sound marketing strategy. Even the world's most famous brands—for example, Coca-Cola—evolved over time, with curated marketing and increased recognition, attributable to their motto and values.

A budget is also required to optimize brand recognition efforts. This is particularly true for digital marketing tactics. Online marketing nearly always needs some form of monetary investment. Fortunately for everyone, these efforts may be tailored to fit any budget.

Promoting your brand through marketing and advertising activities is essential for potential buyers to become aware of it. All marketing communications should encourage brand recognition while emphasizing your *unique selling point*. This is what customers will remember about your organization and what they will begin to associate with it. Customers are more likely to remain with a brand they know and trust, which feelings come from consistent messaging that they know, recognize, and feel connected to.

Constantly maintaining your brand reflects devotion to meeting the demands of your target audience. In addition, it shows that you care about doing so with transparency and effectiveness. With consistent marketing strategies, you can see some significant changes to your overall brand success, and you will see more attention paid to your organization by customers and stakeholders over time.

A well-structured brand outwardly voices the organizational culture and values. This voice should resonate throughout the fire station's media presence, including its website, blogs, social media, and events. You can market your brand in various ways, but you must meet your customers where they are. This means communicating with your customers at their level of engagement and understanding without being patronizing.

Brand management entails setting official style standards for communications. This means consistent messages, logos, and design features. For example, brand standards usually govern which typefaces and colors are used. These rules may also contain editorial directions, such as how to use taglines or specific language including facts and numbers. Brand standards could vary widely, depending on your goals and your fire station.

Your brand provides value to your goods or services. How the resultant pricing compares to a generic version is referred to as *brand equity*. This price differential is supported by product quality and dependability, but with a polished brand, the markup over generic may be substantial. Clients with positive brand equity are prepared to spend extra for identical goods with your brand because they trust and recognize them. A great example of brand equity is Coca-Cola, a globally recognized and internationally sold brand. This iconic company's name has become synonymous with cola drinks, and it is often chosen even when there are numerous other options. This represents high brand equity—even without a high price tag. Thus, Coca-Cola's brand has earned them a top spot in the soda and beverage industry.

Once you've developed your brand, the effort to keep it alive never stops. Marketing is a constant and necessary challenge that organizations of any size and region must face. In all your marketing efforts, your brand rules should be followed. The same tone and feel should be present whether creating logos, taglines, and editorial messages, providing updates via Facebook, X (formerly Twitter), and Instagram, or sending brochures and mailers. In other words, your brand voice should be evident in all you share with your customers, regardless of how and when you communicate.

Finally, all encounters with your brand should promote your brand's identity in a positive way. The goal is to leave your customers remembering your brand and the value that you bring to them, as compared with other competitors.

How Firefighters Are Marketers and Protectors

As a firefighter, you will respond to crises and safeguard people, the environment, and property from various accidents and calamities. You'll collaborate closely with the local community to raise fire safety awareness and prevent fires and accidents from occurring in the first place. You will act and advise on all matters about the protection of life and property from fire and other dangers, including promoting fire safety and enforcing fire safety standards in public and commercial locations. In essence, every firefighter has a role to play when it comes to promoting their brand, demonstrating their value, and even showing the value that the brand has compared to others. How you act and interact with others says everything about your organization, especially at the management level.

At the management level, you might conduct additional supervisory duties such as handling operational issues and supervising the day-to-day work of fire station employees. Managers are also the leaders who customers turn to when they have a concern or question. For managers, their leadership also reflects on the brand and their brand value. While vital, the operational components of firefighting are just a tiny portion of a senior manager's responsibility in a big agency.

Work is frequently performed in hazardous and uncomfortable situations, such as heat and cold, heights, confined spaces, smoke-filled buildings, and all types of weather. You might be at risk from collapsing buildings, cars, explosions, and pollutants. Because firefighters must carry heavy equipment and breathing apparatus, they must be physically fit. This is also a part of the brand value and how the public sees the department in all they do when they serve the community.

People who read the newspaper and online publications, listen to the radio, and watch television are among those you serve and who support you. They may see you in news coverage of fires and other disasters to which you react, but this does not have to be the only time they see you. Seek out your local reporters and make yourself available to act as a fire safety subject matter expert. This will connect your department with the public—your customers—and ensure that your brand value and management consistently remain positive in the community and throughout all your endeavors.

As a result of social media, mass awareness has never been more attainable. Natural disasters and other catastrophes can cause news articles to spread rapidly across the leading social media platforms. Today, firefighters and other first responders are expected to use social media during an emergency to aid in quickly transmitting information. These networks enable us to communicate in a timely and effective manner. Citizens may use social media to get vital real-time updates regarding global catastrophes and other concerns affecting them, as became normal during the COVID-19 pandemic. In addition, citizens may use social media to learn more about the organization, including your efforts and your brand. Social media also facilitates coordination when many parties are involved in disaster response. Of course, firefighters must deal with all these stakeholders, so they must market themselves to provide better services, including on social media wherever necessary and appropriate.

Most people turn to social media for information when a fire incident emerges. Here they receive crucial updates from those in the course of the crisis. The use of social media in CRR plans and disaster management scenarios has prompted the prioritization of social media operations by many response, disaster management, and relief organizations. Through social media, these organizations inform the public about the humanitarian initiatives they are planning. Thus, firefighters must be vigilant in using these channels to communicate efficiently with community stakeholders.

Social media is one of the easiest ways to demonstrate brand value and connect with customers. No matter what platform(s) you pick, the internet can be a great way to communicate with the community, demonstrate value, and manage your brand. Although brand management can be a time-consuming part of your marketing efforts, if done correctly, it can benefit your organization and cultivate relationships with the community.

The Importance of Developing a Relationship with the Community

One core concept in firefighting is community involvement. Specifically, this means includes community people in problem-solving solutions to situations that affect them. The most successful method to achieve public safety goals is to actively involve individuals experiencing the issues in their resolution. Broadly, community involvement includes people of the community in all activities, from identifying essential issues and deciding how to handle them to assessing and sharing the outcomes with the district. The community's people can be anyone from the decision-makers to the stakeholders. As a further benefit, community involvement can also enlighten firefighters about the challenges people in the community are facing and the best solutions.

Community involvement is a core value and practice of public health. The importance of community engagement stems from the notion that the public has an opportunity to participate. The public health community thinks that by pooling our collective knowledge and cooperating, we would be able to detect issues more precisely and design more elegant and effective remedies. Furthermore, if individuals have an opportunity to buy into the process, any potential for conflict will be reduced. The more individuals are involved, the more people will work to resolve conflict where it arises.

As part of community involvement, your organization should bring in leaders from local government, local organizations, and other fire service establishments. By bringing in leaders from community organizations who can and will assist, your organization indicates that it appreciates their contributions and will consider their opinions. This can enhance trust, communication, and receptivity to using services. Community involvement strengthens bonds between individuals, community associations, organizations, and churches, increasing community support. When the community has strong bonds, solutions can be found and problems can be remedied even faster; consequently, the fire service will enjoy a more impactful connection and influence within the community.

Creating and maintaining relationships is at the heart of community organizing. The strength of a community is determined by the power of our ties with one another. People with solid connections can effect meaningful change. Yet, building these connections takes time and work, in the form of consistent outreach, support, and communication between stakeholders and organizations. Moreover, effort must be put into demonstrating the value of the organization and its concern for the community. Despite this, it is well worth the investment, given the impact these relationships can have on our work.

Relations are frequently the source of our greatest joys and our biggest problems. Relationships are not easy to comprehend, having no simple formula since people they comprise are unique and complex. What can work for one community or individual will not work for another, as every relationship must be personalized and a fair amount of effort must be undertaken to create understanding and connection. That said, relationships begin with one simple, expected promise to one another—namely, that respect and kindness will prevail.

We should treat our neighbors how we want to be treated, which is central to nearly every faith. If you keep this golden rule in mind, you will most likely be successful in developing connections on which you can rely. It is also a universal rule to follow in every relationship you build. By following this golden rule when building relationships, you will be able to extend your network further than you could ever imagine to include friends, colleagues, associates, allies, and partners around you, regardless whether in the role of a leader or a citizen. There is no limit to what you can do as a group, especially if you proceed with the right intentions, mindset, and overall mission when making these connections.

When they aren't responding to crises and attempting to save lives and property, firefighters are actively educating others about the significance of fire safety and cardiopulmonary resuscitation (CPR), among many other things. The educational work that firefighters do is among the most important of the community outreach initiatives they pursue. This is critical work because the more upstream interventions are applied to the community, the safer everyone will be. Furthermore, the more the public is educated, the more they can be expected to support the fire service and their initiatives to help save lives; create a positive, uplifting community; and further the influence and role of the fire service in their communities.

How Firefighters Are Perceived in Society

Most people escape and run in the other direction when there are flames and explosions. Firefighters, by contrast, rush toward them. Firefighters are among the first to arrive on the scene when a crisis hits. Even in medical emergencies, firefighters are on the scene. This selfless heroism is why firefighters are so respected and admired.

According to a recent Harris Poll, at least half of everyone considers six jobs of great prestige: firefighters, physicians, nurses, scientists, teachers, and military officers. Of these, firefighters are rated highest.[1]

Firefighters frequently work in high-stress circumstances, especially when individuals are hurt or threatened with danger. Firefighters may be asked to console victims or urge the public to avoid the area when responding to an event. Firefighting is a physically and mentally hard job that requires fast thinking, teamwork, and stamina, especially during major incidents.

Firefighters are brave. They are aware that there is a possibility that things will not go their way. Nonetheless, they persuade themselves to go ahead and do it.

Firefighters are perceived as heroes and protectors as they decipher the true causes of the fires and emergencies plaguing their communities. As firefighters tend to learn on the job, a match does not always spark a fire. Fires can occur when specific substances encounter one another or when an electrical line becomes overheated. Fires are classified into six types; gases, liquids, solids, metals, oils, and electrical. Each necessitates a unique response, and it is up to the firefighters to determine which type of fire they are dealing with.

After extinguishment, firefighters are charged with determining the cause. They use their understanding of fire and logical skills to determine the origin of a fire, both where and when. This involves applying scientific knowledge across multiple subjects, including chemistry, physics, and engineering. To the public, firefighters are heroes because of this unique knowledge base and skill set, which they must understand diagnose, treat, and solve communal crises that can cause property damage and threaten human and even animal lives.

To become a firefighter, you must be a specific type of person. Firefighters jump out of windows, dash through flaming buildings, and pull people from automobiles. They see and experience things that impact them emotionally and physically. Despite this, they keep doing what they're doing.

Being a firefighter entails more than just putting out fires. Firefighters save people. They assist you if you are injured or trapped in a car or structure. They respond to natural disasters and vehicle accidents. In many instances, they serve as emergency medical technicians (EMTs). Firefighters provide medical attention to the ill and injured. They save pets, retrieve animals from trees, and assist the next generation in staying warm.

Firefighters are the great protectors of the community. Those who interact with them believe this wholeheartedly. It's also part of what has influenced the various brand elements of fire service.

Brand Elements and Brand Management

Brand elements are basically the features that a brand uses to differentiate itself from others. Through these, a brand is identified and known by people. If the brand elements are used consistently, it becomes a lot easier to communicate the brand character to the marketplace. Tangible elements of a brand include the product itself, its look, price, and its packaging, among other intangible features. The intangible elements are the experiences the constituents share with the brand and their relationships with it.

As a brand element, the Maltese cross invokes the fire service. Just as the swoosh is iconic to Nike, or the apple with a bite taken out of it is to Apple, the Maltese cross is iconic to the fire service. In many regards, the fire service is America's most trusted brand. As firefighters, we are the marketers, protectors, and ambassadors of that brand. We are given the privilege to serve the community while standing in front of our fire service brand.

The practices of the business world often have implications for the fire service. *Brand management* is a perfect example. Brand management begins with analyzing how a brand is perceived in the market. Then it proceeds to the next step, planning how the brand should be perceived.

Brand management is essential to ensure that the brand's objectives are being achieved and comprehended as prepared while securing its objectives. Firefighters are the brand managers of the whole fire service. From probationary firefighters to the fire chief, everyone can participate in making the service better.

To manage and strengthen its brand, the fire service must develop a relationship with the community. However, to enhance the fire service's brand, we must first understand how our community perceives our organization and the benefits it imparts. Is the brand recognizable within our community? What are our relationships with surrounding agencies? How do we engage and manage our brand within the community? To answer these questions, let's figure out where we stand and the steps necessary to move forward.

The desired end state is positive brand awareness, and this should be achieved by promoting positive attention within our communities. Brand elements that resonate with your community will produce favorable reactions and be enduring.

For example, having a logo with a strong message or a substantial symbolic value will help a fire department to instill a positive image in people's minds. Since people already have a good image of the fire service in their minds, we should preserve it and work toward bettering it. We can develop brand equity through innovation, creativity, and demonstrated value.

The fire services brand is codified in most communities by its mission, vision, and values. These define not only why firefighters serve the community but also how they will do it.

Marketing the Fire Service

In this era, marketing a brand has become a lot easier than before. Since social media platforms have started providing marketing and advertisement services, brands have had incredible marketing and brand management opportunities. Applications such as Facebook, X, Instagram, and Nextdoor offer immediate access to and information about your organization. Opportunities don't end there.

The 21st-century fire service will be characterized by demonstrating value. CRR embodies the fire service culture. Moreover, it can be leveraged as a brand management tool. It is easy to let the great work done to reduce community risk go unnoticed; nonetheless, our leadership is responsible for ensuring that the brand equity of shared community value is recognized and understood. Never miss the opportunity to highlight the fire service's positive impact on the community through social media, as well as other media outlets or community groups. Where and what the fire department focuses on will gain attention from the community.

If the community engages with your organization only through pictures of firefighters on the picket line, endorsing political candidates, or lobbying for wage increases, you're missing the mark. Building community equity and managing the fire service brand requires much more. We need to assume the mantle of the sacred community trust that when a call to service is made from the community, we, the professionals, respond. This means consistent community outreach, marketing beyond critical events, and further communal efforts to demonstrate the organization's brand value. As all significant organizations do, the fire service needs to place its brand in the public eye and instill the brand identity that it wants the community to recognize in each of its messages to the public.

We must also share brand equity with the community. Successes in the fire service are community successes more than they are successes for the individual organization. For example, building or remodeling stations, purchasing new apparatus, upgrading Insurance Services Office (ISO) standing, or achieving an accreditation status from the Center for Public Safety Excellence all demonstrate community value. Never miss the chance to share this market equity with the community: they earned it. By including the public and notifying them of the great efforts and improvements that have been made, you remind them of your role in the community and heighten their feeling of connection with your mission and your brand.

As brand managers and developers of community equity, we need to ensure user-friendly and inclusive opportunities for those belonging to the communities we serve to interact with their firefighters. Open houses, career days, and community events are opportunities to demonstrate the value the community is receiving for the brand they support. These can also be excellent ways to educate the community, share important messages, and recruit new candidates for the fire service. Being involved in the community has numerous benefits, but more important, it allows us to help civilians to understand the power of the fire service and our critical role.

Sharing stories is another way for the fire service brand to connect with the public. In particular, we can share stories where firefighters have helped civilians outside of their primary job. For example, there are many instances where firefighters have taken care of the pets left behind at the scene. As another example, we can show firefighters giving medical assistance to severely injured people so that the first aid is not delayed. Such real-life stories are the ones that people want to hear; if such stories are shared on social media sites, they will have a positive impact on people, and these stories will further broadcast to the public your positive brand identity and your image as community protectors. Through them, people will develop a perception of firefighters as heroes who can save the day single-handedly.

This is the kind of brand management a fire service requires. The marketing of a fire service should be limited to its services and the emotions associated with it. People consider fire services to be the most valuable since they save the lives of individuals in trouble. Therefore, the fire service should be marketed the same way.

Chapter 4 Review Questions

1. Why is it important for the fire service to be involved in the community?
    a. It is not important
    b. To increase profits
    c. To show off their resources
    d. To help civilians understand the critical role of the fire service
    e. To compete with other emergency services

2. How can a budget contribute to optimizing brand recognition efforts?
   a. By decreasing brand visibility
   b. By investing in digital marketing tactics
   c. By focusing solely on traditional marketing methods
   d. By limiting investment in advertising
   e. By ignoring consumer feedback

3. What is community risk reduction?
   a. A term for ignoring community needs
   b. A tool for reducing community engagement
   c. The culture of the fire service
   d. A way to increase community risk

4. How can firefighters promote their brand?
   a. By strictly focus on practical duties
   b. By avoiding interactions with the local community
   c. By raising fire safety awareness and preventing accidents
   d. By minimizing communication with customers
   e. By disregarding the value of the brand

5. How can the fire service enhance its brand awareness?
   a. By ignoring social media platforms
   b. By decreasing community engagement
   c. By having a logo with a strong message or symbolic value
   d. By reducing innovation and creativity

6. The main goal of brand management is to maintain a positive perception among the target audience.
   a. True
   b. False

7. Marketing strategies for the fire service should focus primarily on the tangible services provided rather than the emotions associated with them.
   a. True
   b. False

8. Customers are more likely to stay loyal to a brand they trust and recognize.
   a. True
   b. False

9. Brand recognition can be enhanced through marketing and advertising activities.
   a. True
   b. False

10. Supervisory duties are not part of the responsibilities of management in a fire station.
    a. True
    b. False

11. How does brand consistency in messaging and communication help a fire department to establish credibility and trust with its stakeholders?

12. What role does community involvement play in public health practices and firefighting strategies?

13. What is brand equity, and how does it impact a fire department's pricing and customer perception?

14. In what ways can firefighters use social media to connect with their community and demonstrate brand value?

15. Why is maintaining a well-structured brand voice important for a fire department across various media platforms?

5 THE HIGH COST OF LOW-FIDELITY LEADERSHIP

Fidelity is faithfulness to a person, cause, or belief, demonstrated by continuing loyalty and support. It is one of the most critical elements in harnessing work culture and nurturing long-lasting relationships, personal or professional. Either fidelity is there or it isn't; there is no in between.

Trust is the cornerstone of any successful relationship, personal or professional. Trust is tough to restore in the workplace once it has been broken. According to recent statistics, many employees currently have little faith in their supervisors and senior management.[1]

Fidelity and trust are essential when building lasting relationships in a workplace. Effective leadership requires this specific pair of traits (fig. 5–1).

Low versus High Fidelity

A leader's beliefs and ideals need to be truthful. Trust transforms a random manager into a remembered and respected leader. Gaining the trust of subordinates, so that you can rely on them to complete tasks, is essential for effective leadership. A supervisor can gain their staff's confidence only if they keep their promises to them. Suppose that a worker voices concern to management and the manager assures the employee that the problem will be remedied. In that case, the issue should be resolved long before the employee feels compelled to bring it up again with the manager. Doing so gives employees the sense that they can trust their boss. When leaders demonstrate dedication to their followers, integrity is evident through their actions.

Similarly, a leader might use a variety of measures to guarantee that they have honest values and beliefs or that honesty is at the center of their values and beliefs. Employees must have a favorable relationship with their managers to prosper. Trust is the most critical interpersonal factor for any relationship between persons to thrive. Aside from that, a leader must make fair decisions. We've already shown that an intelligent leader can always distinguish between right and wrong.

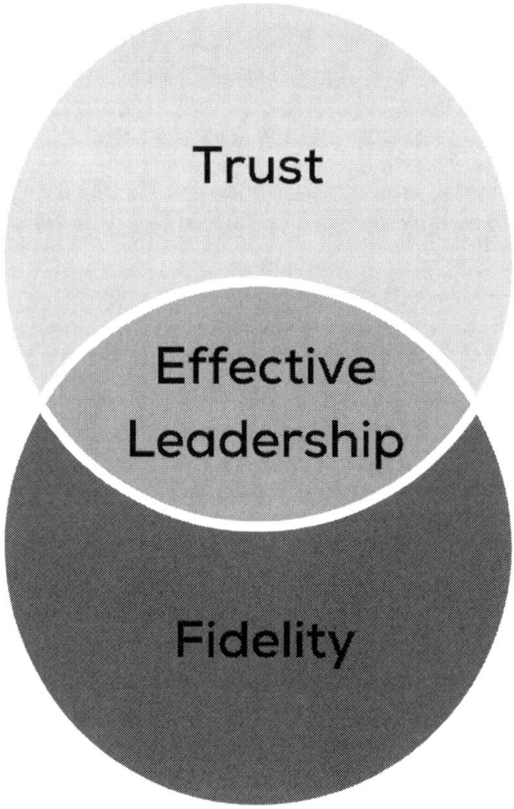

Figure 5–1. Trust and fidelity are the ties that bind an organization together and make effective leaders stand out.

As a result, before taking any action, they should carefully examine the situation and conclude that it looks just and reasonable.

Poor organizational leadership, known as *low-fidelity leadership*, attempts to learn by focusing on mistakes, rather than achievements. However, failure has very low information content, so learning in this model will be limited. Moreover, ineffective leadership comes with a high cost that must be reduced; determining the root causes of low-fidelity leadership usually necessitates either a candid personal or organizational investigation. By contrast, *high-fidelity leadership* is concerned with how tasks are done correctly because success stories are chock-full of helpful information.

Regardless of rank or position within the fire service's typical hierarchical structure, fidelity is a virtue highly sought after. Starting at the top, high-fidelity leaders are the keepers of organizational culture. Many times, practical leadership solutions are required. Their application should happen dynamically across ranks and hierarchies, with leaders, line employees, and corporate members working together.

Furthermore, leaders must recognize and honor their followers' successes to achieve effective leadership. If we're talking about finishing a mission, the commander oversees judgments and provides direction for the troops; the actual deeds, however, are carried out by troops who obeyed their orders. Hence, the commander should not take full credit for the mission's success because it represents a collaborative effort in which each team member contributed equally. Managers regularly and unfairly take credit for their workers' efforts. Followers will perceive that leaders acting in this manner lack integrity and that working under such a leader will always be to their own disadvantage.

Role of Fidelity in the Organizational Operational Model

Fidelity goes both ways; for an employee to be sincere with the work and organization, the leader must put their trust in them. The exchange of mutual feelings breeds trust, transparency, and higher-yield outcomes in any workplace. Unfortunately, I have observed that fire service organizations frequently lack fidelity—and, if somehow present, it is always one sided.

We've already discussed how an employee evaluates the character of their supervisor. Now, let's look at how their leader's honesty impacts workers. When employees feel that their employer is trustworthy, they are likelier to follow orders without question. Only an honest leader can encourage employees to have such faith in them. Healthy relationships between managers and subordinates carry significant benefit to fire stations: When subordinates realize that they can rely on their supervisors, they will approach them with an issue at any time. They may be able to decrease the number of errors in their work and provide high-quality outcomes in less time. As labor productivity increases, the organization's earnings have a better chance of increasing, too.

Employees usually associate a person's integrity with acts of kindness and good intentions. They may be somewhat suitable because a person's ambitions reveal a lot about their nature. Similarly, a fair person who wants to do the right thing for everyone they meet would always want to be kind and pleasant to people around them.

When a leader's actions are based on erroneous goals (i.e., ones that may pose a threat to the fire department), the organization will face significant drawbacks. This will be discussed in more detail in chapter 6 (see "Integrity and Its Impact on Workers"), but the following story conveys the importance of leadership integrity and how it affects the organization and its workers.

A new fire chief takes command of a community's fire department, excited to work with a new agency. The new chief encounters resistance from the start, as the previous chief was beloved and had been with the organization for over a decade. However, the new chief slowly begins to demonstrate their value, showing integrity as a leader in every situation.

When challenging situations are encountered, the new fire chief assures firefighters that they can assist, no matter what. The firefighters begin to notice, many of them seeing the chief's unwavering commitment to them and their department. They are encouraged by the leader's presence, and this translates into the training that they do.

In the first few training sessions, the firefighters are unsure about their new leader, but over time, they exhibit trust in the chief. They now want to emulate the chief's commitment. Most—if not all—of the department's firefighters begin to show up consistently early for work. Many of them put in extra effort. Others volunteer for more community outreaches. Thus, a ripple effect spreads throughout the entire station as people find that the new leader's integrity is rubbing off on them, too.

Under this new leadership, the community fire station begins to reach out more to the community. The firefighters are more motivated than ever, demonstrating the power of leadership integrity and how it can become a true north for any organization, even in the face of new leadership.

As a leader, you must exercise influence. You can make a difference if you stick to core objectives and values that are beneficial and impactful. Influence is an essential leadership characteristic that allows you to shape the actions of one or more individuals or even a large group of people. With influence, you may use your authority to launch new initiatives, make strategic decisions, and effect organizational change. Influential leaders perform activities that others consider critical. Leaders with integrity, such as the fire chief described in the preceding story, can both reinforce the organizational structure and, when needed, transform organizations from the inside out.

> *The glue that holds all relationships together—including the relationship between the leader and the led—is trust, and trust is based on integrity.*
>
> —Brian Tracy

Building Your Organizational Structure

Have you ever thought about what makes an organization's culture successful? Every culture is different, for several factors allow those civilizations to exist. Even despite the diversity in the cultures that exist, certain key attributes must be present for a community to thrive. Incorporating with the six cornerstones of a healthy workplace outlined in the following sections can help you to build a thriving culture.

Environment

Have you ever wondered why some organizations or departments do not have separate offices with soundproof doors? Why do the world's largest financial institutions gather in London and New York City? Or why are Millennials so well liked in online ecosystems? There are several plausible answers to these questions, but the fundamental one is that the environment shapes culture.

An open workplace encourages specific office behaviors, such as teamwork. This can be exceptionally positive in the fire service, where teamwork is a matter of success or failure when on the job—particularly, in crucial situations and disaster responses. Moreover, regional cultures significantly influence corporate culture, either reinforcing or challenging it. No matter where you live in the world, your region will likely substantially impact your work culture. An organization's geographical, architectural, and aesthetic designs, among other (minor) factors, influence employee values and behaviors. The factors that influence an organization's culture are often uncontrollable, but consideration of them—including measurements when appropriate—is essential when developing your organizational structure.

Statements of Vision and Mission

Vision and mission statements are the foundation of a great culture, orienting consumers, suppliers, and other stakeholders to your organization's brand. The simple yet powerful sentences that make up these statements convey an organization's ideas and goals, assisting workers in making day-to-day decisions that lead to their attainment. They establish a brand for the organization when they are vividly portrayed in every encounter between the fire department and its personnel.

Without clear mission and vision statements, your organization may struggle. More important, you may lose direction or labor to complete your desired objectives without your team knowing what they are. Thus, clear vision and mission statements are required to build your organizational structure.

Actions and Rewards

It is critical to put the concepts of your mission statement into action by creating a detailed action plan. If your fire station regards its personnel as its most precious asset, then invest in their education and growth. If your fire department values excellent customer service, then make every effort to ensure that every client, large or small, receives exceptional service. Whatever your vision and mission are, these should directly relate to your actions and what your customers and stakeholders can expect from you when you are on the job.

Establish a reward system to encourage desirable behaviors and actions. Employee performance appraisals, fire department review criteria, and day-to-day operations must all reflect corporate goals. If you do not find ways to reward and recognize the hard work of your staff and the actions that they are completing, then you may struggle to maintain employee morale, and your organizational structure will not be as strong as it could be.

Fundamental Values

After development of your vision and goal, the next step is identification of your core values. Whereas your vision and mission assist in communicating the purpose, core values assist in defining criteria for the actions and attitudes that must be displayed to attain the goal. Having well-defined values conveyed to all personnel results in a standardized vocabulary, which enables the organization to speak the same language internally and externally, to the public. Without clear values, diverse responses and actions can be expected within your organization, for good or ill. Next to your vision and mission statements, your core values are an essential guiding factor of your organizational structure and work that benefits your stakeholders and community.

Storytelling

Each organization has a distinct history. The ability to integrate it into a compelling narrative is critical for building a solid and successful organizational culture. This story's aspects can be formal or informal but become more intriguing when successfully defined, developed, and shared as part of the continuous culture. Similar to proper brand management, telling your organization's story will compel your stakeholders and empower your employees. Whether informal or formal, the storytelling component of your organization will influence your culture and motivate your employees on the job.

People

The crux of any leadership lies in the people it caters to. Creating a culture requires hiring individuals who either share the same beliefs or are willing to adopt those values. As a result, the world's top organizations have strict recruiting standards. They want employees who are not just highly talented but also fit within the organization's current culture.

One study indicated that culturally fit candidates would accept a 7% wage decrease, whereas departments with cultural alignment had a 30% lower turnover rate.[2] As a result, prospective employees choose to work for fire departments that match their beliefs, and current workers continue with cultures they enjoy. Thus, culture strengthens an organization's existing culture.

Leadership Requires Problem Cultivation and Solution Building

"We don't have a North Star," employees complain after leadership teams spend hours creating their fire department's vision, mission, and strategy. Executives are usually taken aback by this feedback, since it implies a lack of aspirational vision; in response, they promptly commit additional time and effort to improve the statement. However, this does little to clarify their people's ideas of a path forward.

Lack of aspirational vision is a prevalent issue that might be caused by multiple variables. Before replying to an employee's request for a road map, consider the reason for their inquiry. You can address the demand more successfully when you understand what's driving it. The next sections detail a few of the root causes of every misunderstanding.

Communication Gap

One major issue that leaders often do not recognize is a communication gap. With employees having various roles, experiences, and backgrounds, the potential exists for failures to communicate clearly, and this can often lead to conflicts, employee disconnect, or even a lack of adherence to the fire department's overall goals, vision, mission, or core values.

Research by Groysberg and Slind in *Harvard Business Review* has found that communication is the root cause of most project failures and conflicts. To bridge this gap, leaders can promote honest, open communication and opportunities for frequent feedback. This will ensure that all stakeholders voice their concerns and understand what the leaders communicate.[3]

Various Elevations

One of the most overlooked issues that can cause misunderstandings is the fire department's various *duties*—or the roles at different levels. While elevations might not seem like an issue at first glance, they can lead to critical issues wherein people fail to understand each other's unique perspectives. From associates to managers to executives, diverse perspectives, values, and priorities impact people's understanding of the solutions and the problems within the fire station.

Communicating to all the various levels of the stakeholders within the department can improve success rates and decrease the frequency of misunderstandings. When people are on the same page and understand each other's perspectives, they feel heard, with their concerns addressed and diverse values accounted for. Therefore, leaders should prioritize addressing and understanding each stakeholder to ensure no miscommunications arise.

Low Fidelity

Low fidelity—or the lack of attention to detail—can cause numerous communication mishaps and often leads to project failure. When stakeholders ignore the elements, the project can easily get derailed and must be recovered or revised entirely. Leaders can increase the fidelity when they focus on receiving detailed feedback from their stakeholders, including creating opportunities for deep discussions and testing solutions. Leaders must understand every detail—including every data point, document, or concern that a stakeholder may have. Without high fidelity, fire departments struggle and miscommunications happen. By contrast, with high fidelity (i.e., exemplary leadership and proper attention to detail), leaders can make decisions that positively impact their fire stations while working toward the objectives of their team and of their organization.

Management behavior reveals the true objectives of an organization. Aspirational statements on paper go in the bin when bosses say yes to everything or fail to consider what their employees say, their employees' priorities, and the concerns at each level of their organization. Employees feel that there is no North Star in this circumstance because, despite a written statement, decisions and individual behaviors are inconsistent with the commitment communicated. This matters when it comes to the fire service as well.

Recall the story earlier in this chapter about the new fire chief and the example that they set. What if they had been the opposite? What impact would this have had on the team and their overall success?

> *Perfection is accomplished not when there is nothing more to add but when there is nothing left to take away.*
>
> —Antoine de Saint-Exupéry

To achieve your objectives and stay true to your purpose, combine your newfound clarity with a decision-making framework specifying the work to be done. Trim the nonessentials and support people who refuse work that contradicts the broader mission so they may focus on what matters most. In doing so, you become your organization's leader while encouraging discipline, another imperative aspect of any successful fire service, as we will discuss in the next chapter.

Chapter 5 Review Questions

1. How can well-defined core values benefit an organization?
    a. They lead to diverse responses and actions.
    b. They promote the organization to use multiple languages.

   c. They assist in defining criteria for actions and attitudes needed to attain goals.
   d. They result in standardized vocabulary only internally.

2. Which of the following is critical for building a solid and successful organizational culture?
   a. High employee turnover rate
   b. Lack of historical narrative
   c. Integration of a compelling narrative
   d. Lack of a distinct story to share

3. How does the chapter suggest influential leaders can transform organizations?
   a. By enforcing strict rules and punishment
   b. By promoting favoritism among employees
   c. By reinforcing organizational values from the inside out
   d. Encouraging conflicts and lack of communications

4. How can a clear vision and mission statement help build an organizational structure?
   a. It creates confusion among the employees.
   b. It restricts the growth of the organization.
   c. It hinders communication within the team.
   d. It assists workers in making day-to-day decisions.
   e. It delays the achievement of organizational goals.

5. Why is it important to have a reward system in place?
   a. To discourage behaviors and actions
   b. To create confusion among employees
   c. To encourage desirable behaviors and actions
   d. To decrease employee morale

6. Integrity in leadership doesn't have any impact on organizational culture or employee behavior.
   a. True
   b. False

7. Vision and mission statements in organizations play a minimal role in guiding day-to-day decisions and interactions.
   a. True
   b. False

8. Bridging communication gaps is not essential for project success in a fire department.
   a. True
   b. False

9. Actions should align with the mission and vision statements to solidify organizational culture.
   a. True
   b. False

10. In the example, the new fire chief demonstrated integrity and value, which ultimately led to increased motivation and commitment among the firefighters.
    a. True
    b. False

11. How does low-fidelity leadership differ from high-fidelity leadership?

12. In what ways can leaders demonstrate integrity in their actions to build trust with employees?

13. Why is it crucial for organizations to recruit individuals who align with their culture?

14. Why is it challenging to restore trust in the workplace once it has been broken?

15. How does a positive organizational culture contribute to the success of a fire department?

6 DISCIPLINE

The backbone of any change is discipline. That little voice within us says, "I need to prepare for the future now. If I want a better life tomorrow, I had better start working on it today." To get your life on the best track, do not wait to work on it tomorrow, but be disciplined about it starting today.

Discipline and accountability (the topic of chapter 7) are the two critical principles of leadership and organizational culture. To reach its full potential, a fire department's organizational model must be centered on these. More generally, if an organization wants to be seen as credible, it must showcase exemplary behavior that evokes a sense of seriousness about being disciplined. Furthermore, a disciplined organization is built from the top down in the fire service. Omitting these from your department is akin to creating a castle in the sand.

Organizational culture directly expresses the organization's demonstrated dedication to its goals, vision, and values. This vision is further translated into employees' productivity and conduct. The mission, vision, and values are not individually exclusive; instead, they are interrelated, and this enables members to traverse a well-modeled organization.

Discipline Defined

What comes to mind when you hear the word *discipline*?

It may evoke a memory of a strict instructor from your childhood or an image of someone famous who has acquired the art of self-discipline. Discipline is a complex concept; it can take many forms, and it is vital from childhood to the end of our lives.

The internet is filled with a plethora of information about discipline in the workplace and its importance. In this context, what exactly do we mean by disciplining the work culture? Although definitions of discipline encompass a variety of fundamental ideas, most include a willingness to obey norms and policies and the expectation of being held to those standards.

Discipline is rarely spoken of in respectful terms in the fire service. In fact, the very notion of organizational discipline elicits a cringe from many. What is the reason behind this? Why does the fire service nearly always regard discipline as an adverse action, rather than a vital means of developing a better organizational culture and employees?

As an organizational construct, discipline is an action taken by a supervisor to alter behaviors or acts inconsistent with the organization's principles, rules, and procedures. The disciplinary process must be established and followed consistently. A supervisor's job is to provide an atmosphere and training that guarantee the competency of all employees. This means that workers should be aware of the performance metrics, conduct, and corporate values expected of them. An employee's willingness to be compliant is determined once they have been deemed competent. Discipline should be applied to those who do not follow the rules.

Discipline is essential throughout our lives and for people worldwide, but have you ever reflected on your own? Self-discipline is crucial for a multitude of reasons. First, it adds to the stability and organization of our lives. We'd be in turmoil if it weren't for self-discipline. Second, it instills in us the importance of accountability and treating people with respect. Finally, every society on the earth is established on rules. Without self-discipline, everyone could do anything they pleased and not care about the consequences. By contrast, having self-discipline in all aspects of life promotes positive human behavior to build a better society and a more pleasant environment in which everyone can live peacefully.

Discipline is not a standalone trait; it brings along multiple aspects of grooming a person's identity and conduct in life. For instance, if your ambition is to be great, then you must first find a way to do so by serving others. If your aim is to be wealthy, then you must first learn to give. If your ambition is to be healthy, then you must first learn to stop doing the things that can make you and others sick. All this is achievable if you stir up some discipline in your daily activities and routine.

You've probably heard people say they wish they could afford to have money left at the end of the month or take a nice vacation to recharge and experience the pleasures of life. Before their lifestyles can change, their wishes need to become a burning desire. You see, if they really desired change, then they would do more than sit at home, watching television every evening idly wishing they were doing or had more. If people really wanted to make their lives better or they really wanted to achieve that goal of theirs, then they would find a way. They would become accountable and disciplined in their efforts to realize their ambitions, as that is the only way to accomplish anything in life.

Why Discipline and Accountability Are Needed in Your Life

If I ever find myself in a scenario where I am unable to perform to my total capacity, I hope I can forget about it and dismiss it as merely a bad dream. Of course, my viewpoint may differ from yours, and my method of dealing with problems may be unique or seem absurd to someone else. However, you get the idea: negative energy cannot be allowed to take over your mind, as it can impact your life.

Negative energy can drain you. To counter it, try to take something away from any negative situation; every one of these has a lesson, something that you can learn from it to improve upon for the future. Furthermore, there is a high possibility that you will face an even more challenging situation in the future. Thus, you should take note of the obstacles in a negative situation, straighten your shoulders, and tackle them head-on.

This sense of acceptance and acknowledgment is derived from *accountability*. Being willing or obligated to accept responsibility for your actions is what accountability entails. Accepting this responsibility means that you recognize the importance of accountability and are committed to making a positive impact. This includes taking responsibility for the decisions that lead to your actions. What makes a good leader stand out is their ability to take total accountability for their choices and their consequences. Good leaders know that they have responsibility at every stage of the journey—from start to finish.

Sometimes, your responsibilities are related to your position in an organization. For example, if you oversee a fire station's finances, you are responsible for being aware of and protecting the fire station's financial situation. If you're in marketing, you are responsible for the turnout of your marketing campaign (i.e., the sales that it produces). As a fire chief, you are responsible for the productivity and efficiency of your station and its staff.

Your responsibilities and your ability to handle them directly relate to your degree of individual accountability. Individual accountability entails accepting and comprehending that your actions have consequences when you oversee certain aspects of your life.

While a common perception of accountability in life is that it means owning up to your mistakes, this falls short of its true meaning. More accurately, it means fully accepting responsibility for the events that occur in your life and realizing that you oversee your relationships, reactions, choices, actions, communication, and attitude. Accountability is recognizing that you are in control of your life, your mindset, your choices, your actions, and therefore your consequences and how you make others feel around you, including your team members.

We are the presidents and CEOs of our lives. Therefore, take control of your life and responsibility for your choices. When you know what you want, don't look back; instead put complete faith in your ability to achieve big goals. A top CEO's actions result in outcomes. A good idea that isn't followed through is a missed opportunity. Opportunities do not come only occasionally, but rather are doors that we choose to open to explore and elevate ourselves to greater heights as individuals.

Whining about your mistakes and faults will not get you anywhere, and it can even work against you. It could become one of your significant worries, preventing you from moving forward. Because of this, you may miss out on various opportunities for personal development. The more you look back, the harder it will be for you to move forward. There are several ways to move past this, but it begins with looking beyond your situation and yourself.

Start providing significant value to others around you, as well as to the work you've been given. Start helping others selflessly. Stop comparing yourself with others; by comparing yourself with others, you waste your time and energy. You are already exceptional, and you don't have to show your self-worth by chasing after others. Not only does this behavior do a disservice to you as an individual, but it also does a disservice to the work and the influence that you could have had if you'd used that energy for good.

Instead, use your time, energy, and resources positively, to bring value to your work (we'll go over this in more detail later in this chapter, in "How to Improve Your Overall Integrity"). Always do the right thing. Do something that shocks you or gives you shivers as you go into it. By venturing outside of your comfort zone, you are free to try new things. Take extra steps to arrive at your goal in a more polished manner. In the end, you will be glad that you went above and beyond—and you will continue to grow your accountability and your discipline in the process.

What Happens When Your Organization Lacks Discipline?

Lack of discipline within the organizational structure detracts from the potential core performance of the organization—in other words, robbing from its productivity. The inability to discipline staff will destroy corporate integrity. Whenever a staff member misses the opportunity to recognize—and learn from—mistakes, the organization, its procedures, and the values that guide the team in their efforts deteriorate. This leads to further deviations from the organization's mission and values, in the form of chaotic behavior where rules and procedures are not followed consistently. As if that weren't terrible enough, lack of discipline undermines corporate leadership's capacity to reflect its ideals truthfully. This isn't fair to the organization or its employees, nor is it acceptable to the people they serve in the community.

Additionally, a lack of discipline and integrity within the workplace leads to employee disengagement. Employees who violate organizational norms or policies must be disciplined to return them to organizational alignment and ultimately enhance their performance. The proper implementation of discipline allows the employee to atone for the violation and will enable them to regain their credibility. Without acknowledgment of mistakes and proper guidance about how to improve, employees become disenchanted; their disengagement then leads to even further dilemmas that can impact the organization's overall success and productivity. Fortunately, this all comes down to leadership—and ensuring that the proper leader guides the team.

A resonant organizational leader (as discussed later in chapter 13) acknowledges, corrects, and coaches team members to return to organizational alignment and resolves competency versus compliance issues. Positive steps should be taken to ensure that team members feel that they are a significant part of the effort and contributors to the overall goals. Employees facing discipline should also be given a chance to make the situation right and learn from their mistakes with the guidance of a credible leader. Supervisors who neglect disciplinary action deny individuals and the organization the opportunity to progress.

Leadership is paramount to the organization's discipline and overall accountability. Without an accountable, positive, disciplined leader, the organization may fail to thrive or to reach its full potential.

Disciplined versus Nondisciplined Employees

Let's consider a practical challenge. Consider two employees: one who received discipline and was allowed to atone for their wrongdoings; and a second, whose experience was the opposite. The first employee, with the chance to correct their mistakes, will most likely be viewed in higher regard by their colleagues than the second employee, burdened with an unpaid debt.

When employees who are competent and compliant notice a lack of required discipline, the message is that obeying the rules is voluntary, not obligatory. This disrupts the momentum of the organization's model. If employees see examples such conveying that they do not have to be fully engaged with the organization's efforts, then team members will likely participate at their convenience; consequently, those who participate fully will become bitter over the dichotomy that has developed over time between those who participate and obey rules and those who don't.

The perception of a double standard harms organizational morale and credibility. Imagine what would happen if a few employees choose when to participate and follow procedures while others adhere to the requirements: An additional burden would be placed on the latter group, as they would have to make up for the mistakes of their counterparts. It would not be a cheerful sight nor a culture that anyone would want to be a part of. As mentioned earlier, such situations lead to lack of integrity within the workplace, which can have dire consequences.

Integrity and Its Impact on Workers

We have discussed how an employee determines the character of their leader. Now let's detail how workers are affected by their leader's integrity. When employees can both believe that a manager is trustworthy and themselves be trusted, they will blindly follow instructions. Only with honesty in a leader would subordinates develop such faith in their manager.

Achieving a positive relationship between the employees and their manager significantly benefits the organization. When subordinates can rely on their manager, they do not hesitate to approach and bring issues to the manager in a timely manner. This way, employees can minimize the errors in their work to produce a quality outcome in less time. Moreover, the organization can maximize its profits as the efficiency of its workforce increases.

Employees associate an individual's acts of kindness and positive intentions with their integrity. They might be right to do so, as a person's goals reveal a lot about that person's character. Similarly, a fair person who prefers to always do right by everyone will be kind to the people around them.

When a leader, through their actions, demonstrates inappropriate intentions (i.e., that may pose a threat to the organization), two different outcomes are possible. One will be that the workers who realize the manager's act is wrong will lose their trust in that supervisor and find it challenging to work under them. Second, those individuals in the organization who do not care about the fire station's well-being, instead looking out for their best interests, will feel more comfortable doing something wrong, given that their manager does the same. In either case, the organization faces fatal disadvantages.

Now, let's consider a short story that encapsulates the importance of integrity in a leader. This will give you a clearer picture of how a leader's integrity impacts the organization and its workers.

Jeremy Thompson has been with a small manufacturing company for several years as a marketing executive, but he has not always been the most successful marketing leader. In years past, many of their marketing and advertising campaigns struggled, and Jeremy's employees failed to meet their objectives or complete projects on time. This led to higher turnover rates, but Jeremy's team were not held to blame.

Jeremy's leadership was questionable given several vital issues colleagues noticed. He was guilty of not soliciting employee input and insight when making many pertinent decisions for his department. Jeremy was a kind boss during meetings, but many employees reported otherwise having difficulty reaching him during regular work hours. Jeremy did not practice an open-door policy even though he stated that he had one, although he did return emails quickly. Consequently, he often left his employees dealing with challenges on their own until crises had emerged. These crises were difficult to manage, leading to meetings to discuss what went wrong—and employees often expressing behind closed doors their desire to leave because of stress.

Furthermore, Jeremy's leadership led to a reduction in employee morale and made employee retention a challenge. Jeremy's marketing division had the highest turnover rate among all the divisions at their small manufacturing company. In exit interviews, outgoing employees frequently indicated that the leadership was not always available, leaving them feeling like they had to endure challenges independently. Given these issues, Jeremy's leadership was worse than that of his counterparts in other divisions. Yet, Jeremy's tactics had led the department to capture more corporate clients and partnerships than ever before.

When Jeremy took a leave of absence to handle a personal matter, his second in command, Julia Smith, took over. Julia had more experience in human resources, and although she had been with the company for about half as long as Jeremy, she knew the company's policies and procedures. Jeremy would be gone for at least two months, during which Julia would keep the department running as usual.

On the first day in her interim role, Julia held a team meeting where she promised she would be available to them. To keep up to date, she scheduled more meetings. She would have a true open-door policy, so her subordinates could bring issues to her before crisis points. Many employees scoffed, remembering how Jeremy had promised the same. However, Julia ended up surpassing the staff's expectations: she followed through on every promise, which made for not only more successful projects but also happier employees. Furthermore, she brought up essential decisions with the team before making them. This was in stark contrast with Jeremy's approach, which many employees had felt was distant.

In the end, Julia conquered the one issue that had plagued the department for years under Jeremy—namely, retention rates. Many employees who had attempted to transfer to a different department had removed their applications. By the time Jeremy returned to the company, a formal meeting was set up to decide what should happen. Julia had impressed higher-ups with her record, so much so that Jeremy was asked to clear out his office and Julia was promoted. Stunned, Jeremy left the company, after which the marketing department thrived, meeting new goals and setting new retention records for the marketing department. Julia and Jeremy, while similar in their kindness to employees, exhibited different leadership tactics, and only one was ultimately successful.

There was a clear difference in the leadership that Julia and Jeremy presented, and it comes down to one factor: integrity. Julia chose to adhere to the values she stated, the ones she knew guided the company. By contrast, Jeremy was lazy in his relationship building, leading to great metrics but poor employee morale and retention rates for the department.

As you can see from this story, it is clear when a leader has integrity. This is even more apparent in the results that you achieve.

How to Improve Your Overall Integrity

To improve your overall integrity, you must actively take a few steps. At least begin somewhere. Take the first step, no matter how small or in which order.

One possible first step is to develop disciplined habits. We all have unlimited potential. We need to start believing in this potential and build on it—to add value by going the extra mile. A *value-creation strategy* is about how you create, capture, and sustain value. This will then translate into the way that you lead your staff, as well as how much you all succeed together.

Second, you must not only hold yourself accountable as an individual but also hold others accountable for their actions. Accountability is one of the most important aspects of a healthy and fulfilling life because it leads to more beneficial relationships, characterized by honor, compassion, truthfulness, dedication, and sincerity. Without accountability, these relationships suffer, often putting you in situations where they do not provide a positive impact, or you fail to have meaningful relationships at all. In the workplace, discipline and accountability make for equitable relationships and respect for one another, and they also encompass the crucial elements of all leadership theories.

Discipline and Accountability Are Critical to Every Leadership Model

The leadership theories explored in the chapter 1 are nothing without discipline and accountability. A leader—whether transformative or traditional in scope and practice—cannot be a true leader without accountability and discipline, as there will be no model for how this is meant to show up in the workplace. In adaptive leadership, employees cannot follow the leader if they are not accountable for their choices, even in the face of transformative changes. Servant leaders require accountability and discipline to show up daily for their team; if they are absent, employees will notice and the team will struggle.

Discipline and accountability remain critical to every leadership theory for various reasons. First, they promote the achievement of goals. With accountability and discipline, leaders guide their teams to success through clear goals, directions, and a pathway. These two attributes make their framework for success simple—essentially ensure that leaders and team members take responsibility for their progress and development along the way.

Second, leaders who demonstrate accountability and discipline remain trustworthy and credible. Imagine if a leader did not possess either of these traits. How could you follow them to success if they were not accountable and did not hold to their promises? Would you even consider it? That's why accountable, disciplined leaders succeed more: their teams trust them, and their employees buy into their leadership faster. More important, the more trust and credibility that a leader fosters, the more inspired employees will be to follow the leader's example.

When employees follow their leader, this also drives organizations to improve their performance and productivity. If everyone remains on task and accountable like the leader, then objectives are met, tasks are completed, and organizations see a greater degree of performance and productivity that enables them to elevate their team's successes. Incorporating discipline and accountability throughout the leadership and the entire organization also encourages continuous improvement, which allows the team to succeed when they find areas of improvement and leverage their strengths.

Evidently, without leadership that exemplifies discipline and accountability, the results never lead to this growth and excellence. This also leads to employees who fail to hold themselves accountable because of lack of discipline—which is possible regardless whether leadership exemplifies these characteristics. Unfortunately, as we will discuss next, not all fire stations have the discipline and accountability they need to succeed.

Why Do Many Organizations Fail to Have Disciplined, Accountable Employees?

This question continues to puzzle those within teams and those who study leadership. Ultimately, there is no single answer for every organization. Some people might say such failure is the leader's fault, while others might blame different personalities within the workplace. The answer about what is at fault or missing is elusive. To illustrate this, if you were to ask me what darkness is, I would have to answer in terms of its opposite, light: dark is the absence of light; in other words, there is no definitive description of dark on its own. Similarly, you cannot give a one-line response that fully and accurately conveys why many organizations fail to have disciplined and accountable employees.

The answer to why employees lack discipline or accountability requires understanding of where there is or is not discipline within other areas of the organization, such as with the leaders themselves. Multiple factors build up to such devastation within an organizational structure. To restore discipline and nurture an environment where employees take responsibility for their actions, a leader needs to address the root cause of the problem and lead by example.

A leader must overcome the habit of being insecure in the workplace, believing that it would benefit someone else rather than themselves, because it is evident that the benefit of any team member would help the entire workforce and the organization when leadership is involved. Instead of allowing insecurity to govern, leaders must engage in *vision casting*, which means communicating their vision with others. As real-world examples of vision casting, the organization's strategic landscape will shift if the market strategy shifts; and as the agency expands, you must adjust your vision and present it to the organization along with completely new goals to work toward.

A leader is accountable for everything that happens at work, which means that if you or a team member makes a significant mistake, it is your role as a leader to accept responsibility rather than pointing fingers or allowing other team members to criticize their peers. As the leader of the team, you are responsible for all the team does, even if you did not directly do it yourself. This is not to absolve someone of blame but rather to set an example.

Criticizing and shaming other team members not only affects the person you're publicly blaming and humiliating in the workplace but also fosters an unhealthy environment in which individuals are frightened to make a mistake. This is wrong because it denies a person's right to feel protected and prevents them from learning from their mistakes and avoiding them in the future. In a team environment, there should always be room for errors, and employees should feel empowered once you guide them to learn through their mistakes.

As a leader, you must instill a courageous culture in your organization. Encourage people to speak up and share their thoughts and concerns at work. Employees who feel heard and responded to are far more productive and innovative. If your employees feel that you care about their perspectives, then they are more likely to be involved in the team and become even more committed. After all, being heard by their leadership will further increase their feeling of inclusion in the workplace.

Even if you are not personally at fault when a mistake occurs, it is vital that, as a leader, you accept blame and take corrective measures. Extraordinary leaders recognize when it is necessary to take responsibility. They then work hard to remedy them while avoiding unnecessary drama.

Regardless of whether you or a team member makes a mistake, your approach to resolving the problem should be the same. As a leader, you must accept responsibility for both the good and the bad to keep the workplace in balance.

It bears repeating that a leader must lead by example, blazing a trail that is easy for others to follow. Once you master your leadership skills and demonstrate the pathway to subordinates, they will follow you. When your team is valued, they are loyal and will value you as a leader. If they are not respected, though, they lose their connection and may not be as dedicated to the team effort, which is paramount in the fire service because you all serve the community.

Being the leader, if an endeavor fails, you are rightfully meant to be held accountable and answerable. Once you have taken the lead, it is your duty, more than just a goal, to demonstrate your proficiency and expertise in the capacity by giving directions and providing recommendations for people who encounter predicaments in completing their tasks.

Worldwide, the digital workforce is expanding and offering opportunities, and this tendency is expected to continue in the following years. Technology is reducing barriers, allowing possibilities to emerge that do not recognize previous boundaries, for a new borderless reality. The only difference is the scope of your ambitions and creativity.

In the United States, 76 million people work part time from home, or approximately half of the total workforce. Employee retention is a significant issue for any fire station, particularly among younger generations. This will substantially impact office design, travel, and real estate enterprises.

In the past, employers hesitated to embrace remote work or work from home because of a lack of confidence in their workers' accountability. However, data show that those who were compelled to work from home during the COVID-19 pandemic had higher productivity. This new world after the pandemic presents incredible opportunities for anyone willing to take on a challenge and put their skills to the test. Simultaneously, it puts up enormous obstacles, the most difficulty being maintaining employee accountability and setting appropriate expectations for organizations and departments; a delicate balance must be struck between the two.

Discipline Is an Opportunity!

Disciplined employees can atone for their conduct and return to alignment with the organization's values. Organizational culture and long-term health depend on discipline and the atonement process that follows. Strong leadership is essential to realize the potential to impose discipline. Don't pass up the chance to be disciplined.

> *Discipline is the force that prompts individuals or groups to observe rules, regulations, standards, and procedures deemed necessary for an organization.*
>
> —Richard D. Calhoon

To have discipline is to systematically conduct organizational activities while strictly abiding by the rules and regulations of the organization. For any organization—not just the fire department—to achieve the optimum level of productivity with positive outcomes, the organization and the employees need to work coherently with each other. In other words, the organizational members must work as a team to achieve the organization's objective and vision, recognizing that individual and group goals and wants must be aligned for the organization to succeed.

An organized employee will always have discipline, and a disciplined employee will always be organized. Employee behavior is the foundation of organizational discipline. Discipline entails adhering to the organization's specified code of behavior. Discipline ensures productivity and efficiency. It promotes employee harmony and cooperation and provides a boost to employee morale.

An organization will experience turmoil, confusion, corruption, and disobedience when there discipline is lacking. This is because organizational discipline entails employee obedience, orderliness, and proper subordination. Organizational discipline is aided by work recognition, the fair and equitable treatment of employees, adequate compensation structure, competent grievance management, and job security.

Objectives to Achieve by Instilling Discipline in Your Workplace

When you instill discipline in the workplace, you can easily achieve numerous objectives. For instance, you can devise a uniform system where everyone follows the rules and regulations while following the organization's work policies to achieve timely goals. Without discipline, it is impossible to perform tasks together, and it will be even harder to ensure that the job is completed on time. With discipline within your organization, you can enjoy a uniform system that guides every member—including you, as the leader—to achieve goals together and by their deadlines.

Discipline is also necessary to establish an atmosphere where your team values transparency, honesty, loyalty, integrity, and tolerance. A nurturing, growth-oriented environment will feature all these qualities that make for exceptional team members while prioritizing open communication and commitment to the team's goals. You and your employees will see a shift in your organization consistent with the values that you emphasize. Ultimately, this will lead to tremendous success for you and your entire fire department.

A nurturing environment guides your team to create a culture where opinions are valued and individuality is respected. Your team will prioritize the correct values and remain disciplined, when the culture is positive and your team members feel that they matter and are heard. This vital objective ultimately works to achieve everyone's most important collective goal: to channel your efforts in the right direction for tremendous success!

Once you have curated the perfect employee culture and have emphasized the correct values, your team could be unstoppable, with unity and direction. From there, your objective should be to continue this forward momentum, to improve it, and to watch what happens. This can lead to instant success for your team and your entire organization.

Discipline is a constructive attribute that has tangible results. It establishes a work culture that encompasses all the aforementioned aspects of a successful organization to increase employees' morale and work efficiency.

Discipline is essential for the smooth running of an organization and for the maintenance of workplace productivity, which is the foundation of industrial democracy. Without discipline, no enterprise would prosper.

Embracing Discipline to Protect the Organization

Discipline is both challenging and rewarding. Emotionally sophisticated people must navigate its application. Furthermore, healthy organizations welcome the opportunity to discipline so that they may defend the organization and employee's worth and integrity. Encourage discipline as a means of preventing these from being lost.

The supreme quality for leadership is unquestionably integrity. Without it, no real success is possible, no matter whether it is on a section gang, a football field, in an army, or in an office.

—Dwight D. Eisenhower

One essential quality of a successful leader is integrity. A leader is a person who shows the right path to a group of people and guides them to their destination. To do so, it is necessary that first, the leader has a clear vision of that path. A leader is able to tell the difference between what is right and what is not only through their quality of integrity.

Any organization—whether entrenched in the corporate world or providing social services to the public—seeking a prospective leader will look for an individual who can differentiate between right and wrong and has the strength to then do what is right. In the fire service, when a person is hired for a manager or supervisor position, that individual must possess such integrity. This is because a person who is honest will automatically prioritize the fire station's policy; they will follow all the rules and regulations when working in that organizational environment; and personnel working under them will also start honoring the department's policy if they aren't already. Hiring a leader with integrity helps an organization to develop a positive workplace culture and enhance the performance of the human resources making up their workforce.

As discussed in chapter 5, a leader bears great responsibility. It is true that when an individual gains a designation in a fire station which requires leading a particular group of people, they must be cautious about their skills and competencies. Fire departments value that person the most, showing that they care about rules and ethical behavior. Many professionals from the corporate world believe that leaders who value integrity are a great asset to the organization.

Tim Hird, a well-known professional in the corporate world, shares these thoughts about the impact of integrity in a leader's character on a company's operations: "Leaders who act with integrity and treat people well help maximize the contributions and build goodwill for their organization. As important as these attributes are for managing a team, they also drive business by attracting investors, customers, and potential staff members."[1] The company first needs to ensure that the people who sit in its higher management value honesty and integrity. Then, when selecting an operations manager, for example, they check for these traits in the candidate. This process can be considered as a complete chain of actions.

Moreover, change starts from the top: the person who falls in the lowest level of the hierarchy in a fire station always follows the footsteps of higher authorities. When followers see their leader caring about doing the right thing and honoring the organization's profitability, the employees will most likely adopt the same practices. As a result, an overall change is observed in the organization's environment, which satisfies the organization's stakeholders and attracts new professionals who value ethics.

Fire service leaders need to realize that they are shaping the fire station's culture and values. Their decisions, management techniques, and actions define the organization's culture. At the start of this chapter, we learned that managers need to build the trust of their subordinates to work with them effectively. When leaders understand that their followers determine the effectiveness of their decisions based on their integrity, they will try to treat their followers in the best possible manner, internalizing this tendency toward kindness such that it becomes automatic. This will not only create a good relationship between the managers but also bring a positive change in the workplace, where workers and supervisors both respect each other's judgment and opinions.

Every organization wants good leadership and requires that its employees conform to the necessary system prescribed by the fire department's motto and vision to achieve desired goals. Thus, to summarize the whole concept, an organization expects its employees to show discipline through orderly conduct. This is to encourage not just one errant employee but all the other employees at work as well to behave sensibly to keep the workflow running smoothly without disrupting the momentum. This will keep employees mindful of the rules and regulations and what they must achieve unanimously as a team, ultimately creating more accountable employees in even the most chaotic situations.

Chapter 6 Review Questions

1. Why is it important for a leader to accept blame and strive toward corrective measures, even if they are not personally at fault?
    a. To avoid responsibility
    b. To cause unnecessary drama
    c. To increase employee loyalty
    d. To demonstrate proficiency and expertise
    e. To maintain workplace balance

2. What is the impact of lack of discipline within an organization?
    a. Boosts core performance
    b. Strengthens corporate integrity
    c. Results in chaotic behavior
    d. Ensures consistent adherence to rules

3. What is the importance of learning from negative situations and tackling obstacles head-on?
    a. It avoids all future challenges.
    b. It allows continuous improvement.
    c. It leads to comparing oneself more to others.
    d. It encourages whining about mistakes.

4. How does providing value to others contribute to personal development?
    a. It results in isolation from others.
    b. It encourages self-centered behavior.
    c. It fosters a positive impact on others and oneself.
    d. It leads to decreased accountability.
    e. It promotes constant comparison with others.

5. How do disciplined employees differ from nondisciplined employees in terms of organizational regard?
    a. Disciplined employees are viewed less favorably by their colleagues.
    b. Nondisciplined employees lead to greater organizational success.
    c. Disciplined employees are regarded more positively by their colleagues.
    d. Nondisciplined employees are more compliant and competent.

6. Accountability is about recognizing that you are in control of your life and your actions.
    a. True
    b. False

7. Integrity is not considered an essential quality for a successful leader.
    a. True
    b. False

8. Subordinates are likely to trust and approach their managers more if the managers exhibit dishonest behavior.
    a. True
    b. False

9. A good leader stands out by taking total accountability for their choices and the consequences.
    a. True
    b. False

10. Leadership exemplifying discipline and accountability may not impact employees' behavior.
    a. True
    b. False

11. What key qualities do leaders need to demonstrate to create a successful and ethical work environment?

12. What are the consequences of a lack of discipline within an organization?

13. How can good leadership favorably influence employee behavior and promote a positive work environment?

14. What does it mean to be the president and CEO of your life in terms of accountability?

15. How does providing employees with opportunities to atone for their mistakes contribute to a positive organizational culture and climate?

7 ACCOUNTABILITY: THE ANTIDOTE TO CHAOS

Accountability is the glue that ties commitment to results.

—Bob Proctor

Accountability is a cornerstone of successful fire service organizations. Neither rank nor title confers accountability—not by a long shot! Accountability is taking ownership of one's actions and, in turn, their consequences. It is not applied merely when convenient or expedient. Consistent and calculated actions and efforts define accountable leadership.

Employees who witness or directly experience a lack of accountability will hamper employee engagement, participation, and ownership. It is not the leader but the lead that will bestow the accountability credential. Committing to the organizational values and mission means being committed to and responsible for your actions.

As the leader of an organization, you ensure that the right people are assigned to the right task. This sends an unmistakable message about how serious you are about the job and, more important, how the job will turn out. Not being accountable for your actions doesn't make you a good person. As the aphorism says, "The road to hell is paved with good intentions." This talks about the responsibility that comes with taking the lead. Personal and organizational accountability keeps the organization's goals and values safe.

Normalization of Deviation

How do organizations depart from a culture of accountability? Essentially, struggling managers and leaders allow what they encourage. Lethargy can spread through the ranks when organizations do not hold their employees—and themselves—accountable. As result, some employees' performance will deteriorate to the lowest "acceptable" level that management tolerate in this environment.

As an implicit requirement from their employer, employees discreetly demand organizational accountability. People produce their best work if they believe their performance will be noticed; therefore, being held accountable has a beneficial outcome. Demanding accountability from subordinates but failing to be responsible yourself should be the punch line of a poor joke.

The insidious effect of failing to hold individuals and organizations accountable is the *normalization of deviation* from high standards. The rank and file become numb as the range of acceptable behaviors increases and the performance threshold gradually decreases. Employees, particularly subordinates, should know when they are held accountable and, more important, when not. Personal investment and engagement in the organization's goals and culture lead to the development of a culture of accountability as the antidote to chaos. When accountability becomes an organizational value, the need—and desire—for micromanagement will fade.

Organizational Freelancing

A leadership ethos that does not value accountability will influence peers and subordinates to do the same. Deficient accountability creates a permissive organizational backwater where individuals can do whatever they want, similar to freelancing on the fireground. When action or inaction reinforces the perception that people can behave and act without consequence, *organizational freelancing* becomes the expected behavior, making accountability an afterthought.

Exercising Accountability and the Accountability "P"

Accountability is not a single reference point but rather a continuum. The National Incident Management System illustrates the planning cycle process through the Planning "P"; this is used for incident accident planning in emergency and nonemergent operations. Let's look at accountability similarly by redrafting the Planning "P" as an *Accountability "P"* (fig. 7–1).[1]

In the Accountability "P," the long stem of the letter "P" represents the introduction to the opportunity or challenge to demonstrate accountability. That is then evaluated and measured against one's leadership ethos and the organization's mission, vision, and values. Engaging in the accountability process begins at the decision point with a desire to act. The loop of the "P" then represents one's engagement through the decision-making process, which requires clear, concise communication and careful navigation of the organizational culture by the leader.

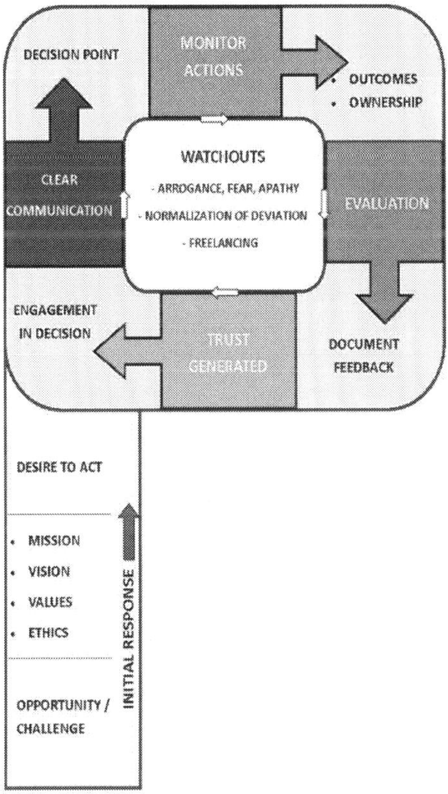

Figure 7–1. While accountability within an organizational leadership structure is vital, it is not easy. This Accountability "P" will allow you to assess how well you and your organization are demonstrating and reinforcing a culture of accountability.

Attitudes and behaviors that influence accountability cannot be underestimated. We have reached the decision point through consideration of them. Be decisive and invest incredible effort. After the decision point, one must endure the outcomes and take responsibility for the decisions. This is the crux of accountability. When taking responsibility for the outcome—good, bad, or otherwise—you must own it. Praise favorable outcomes and take responsibility for adverse ones.

Now we can transition to the evaluation phase of our Accountability "P." What worked, and what could we do better next time? Do outcomes support our organizational mission, vision, and values? Do we own the outcome and decisions associated with the results? If so, we are demonstrating accountability. This process, repeated over and over, is what generates trust and fidelity with our organization's membership. Documenting what worked and didn't work is critical to effect change. Asking for feedback is an emotionally intelligent response and produces tangible dividends in our relationships with others. We are now ready for the next challenge as a new opportunity to enter our Accountability "P" continuum.

Leaders who believe they do not need to be accountable irrationally assume that accountability is only for subordinates. They feel that applying supportive or corrective action and feedback to every employee's decision would be a useless and annoying practice. This is not a tenable leadership paradigm.

When leaders hold employees accountable for their actions and provide feedback and remedial actions, this shows that their actions and behaviors impact the organization's mission. Take servant leadership, for instance. Servant leaders invest their time and energy into giving their employees the feedback they need to understand their situation and their mistakes and become better people. Regardless of their alignment with this methodology, leaders should employ a similar mindset to continually invest in and support their employees and teams.

Investing in employee accountability will create an engaging culture where personal ownership of performance outcomes can be celebrated. If not addressed promptly and regularly, arrogance, fear, and indifference will hinder employees' desire to be accountable. Move forth with confidence in your ability to be responsible, which is a requirement. Accountability builds corporate credibility and is perhaps trust's most valuable commodity.

Accountability and Constructive Feedback

Although we are discussing the leadership changes that can be brought within the firefighting department, these parameters can be implemented in any organization. For the growth and development, it's crucial that the employees' skills and knowledge continually improve. If employees are not made aware of imperfections, they will never improve. Therefore, it's also vital for a business owner to frequently assess the performance of the employees and provide them with regular feedback.

Through periodic evaluations and feedback, your team will be able to identify their mistakes and learn how to improve the quality of their work. Importantly, make sure that the remarks you give to the employees are constructive and not destructive, so that they learn something, rather than becoming demotivated.

Suppose you want to see a positive change and development in your employees. In that case, you need to appreciate even their small efforts and comment negatively on their performance comparatively less. The best way to enhance the efficiency of your employees is to adopt the *carrot-and-stick* approach, which has nothing to do with actual carrots or rabbits. Carrots are the rewards you give your employees, and sticks are the penalties you impose when they do something wrong.

A resonant organizational leader recognizes and corrects issues of competence versus compliance and coaches members back to organizational alignment. Proper discipline allows the employee to atone for the infraction and provides an opportunity to restore their credibility. If supervisors fail to address actions worthy of discipline, they are failing the employee and depriving the organization the opportunity to grow. A good leader is expected to create a healthy and effective work environment in which employees are engaged and empowered.

Providing constructive feedback is a vital skill for leaders, fostering growth and improvement within their teams. Effective feedback should be specific, clear, empathetic, and solution oriented. Next, we'll look at how leaders can provide constructive feedback, focusing on making observations, being direct, showing empathy, and providing guidance.

Observations

The first step in providing constructive feedback is making precise observations. Leaders should focus on specific behaviors or actions, rather than generalities. Observing and noting specific instances allows the feedback to be grounded in concrete examples, which makes it more actionable and credible.

Tips for Making Effective Observations

- *Be objective.* Stick to what you saw or heard, avoiding assumptions or interpretations.
- *Be specific.* Mention particular instances, rather than vague generalizations.
- *Be timely.* Provide feedback close to the event, when the behavior is still fresh in everyone's minds.

For example, instead of saying, "You're always late," an effective leader might say, "I noticed that you were 15 minutes late to the last three meetings."

Being Direct

Once observations are made, the next step is to communicate them directly. Being direct does not mean being harsh; rather, it involves clear communication. This clarity helps the recipient understand precisely what is being addressed, without ambiguity, and why it is important.

Tips for Being Direct

- *Use clear language.* Avoid euphemisms or beating around the bush.
- *Stay focused.* Keep the feedback focused on the issue at hand.
- *Be honest.* Provide an honest assessment, even if it might be uncomfortable.

For instance, instead of saying, "Maybe you could try to be more punctual," say, "It's important for our team that everyone is on time for meetings."

Showing Empathy

Empathy is crucial in delivering feedback constructively. It builds a positive rapport and ensures that the feedback is received as intended, rather than as a personal attack. Empathetic leaders are attentive to the feelings and perspectives of their team members.

Tips for Showing Empathy

- *Acknowledge* feelings. Recognize how the feedback might make the person feel.
- *Be supportive.* Show that you are there to help and support their growth.
- *Listen actively.* Allow the recipient to share their side and listen without interrupting.

For example, an effective leader might say, "I understand that time management can be challenging, and I appreciate your efforts. Let's find a way to ensure that meeting times work better for you."

Providing Guidance

Constructive feedback should always include guidance on how to improve. Simply pointing out what is wrong is not enough; leaders must help their team members to understand how to correct their behavior or improve their performance.

Tips for Providing Guidance

- *Offer solutions.* Provide specific advice or strategies for improvement.
- *Set goals.* Help determine clear, achievable goals to guide improvement.
- *Follow up.* Check in periodically to offer further support and acknowledge progress.

An effective leader might say, "To improve punctuality, perhaps you could set reminders for meetings 10 minutes in advance. Let's review how this works in our next one-on-one meeting."

Effective feedback is a blend of clear observation, direct communication, empathetic understanding, and practical guidance. Leaders who master these elements can foster an environment of continuous improvement and positive development within their teams. By being objective in their observations, direct in their communication, empathetic in their delivery, and supportive in their guidance, leaders can provide feedback that not only addresses current issues but also encourages growth and development for the future.

An Organizational Leader's Responsibility

Leadership is not a position; it's more of an action, a complete code of conduct. You must align your thinking with your subordinates' to run organizational operations smoothly. As Heath asserts in *Upstream*, you can shift your mindset and perspective and further align with your associates for increased success.[1] The fundamental principles of an organization and their alignment can increase engagement and productivity.

You become a leader when your actions inspire others to dream, learn, do more, and be better than they were yesterday. When you envision employees with what you perceive to be the ideal traits, employees understand the fire department's purpose and adopt it as their own. In every organizational process, this practice aligns everyone's efforts and directs their energies into achieving mutual goals as a team.

People are the crux of every leadership effort. Leaders cannot lead until and unless they understand the people they are leading. Every transformative leadership framework requires that leaders look beyond themselves to see their stakeholders and their employees, their needs, and how to meet these to support the team as they progress toward their objectives.

A Unified Goal

One way to look at leadership is that the function of a leader is to lead and guide people who will follow the same values. Therefore, an effective leader must be able to build relationships and create trust and honesty to lead the team toward a unified goal.

A true team whose members work together by directing their minds, bodies, and energy in a singular dimension has a unified vision. By contrast, a collection of individuals with distinctive qualities working in multiple dimensions does not make up a good team. Nor do people suddenly transform into a group merely because they work in the same organization. Vision is the essential factor uniting the group. Any fire department can embark on any venture easily if they have unity and faith in each other, the organization, and the management.

Unity begins with the leader. You must be vivid in your vision and tactics in achieving the goal to bring the team together. Many organizations suffer because the people in leadership positions tend to have no clear vision, or maybe they have a vision but do not know how to execute it. Even though it is often assumed that the staff know and understand what is happening within the fire department and the purpose behind a given task, this is not always the case. Lack of vision ultimately creates a severe threat to the organizational structure and operations, with people just following the rules without proper instructions and lacking any clear sense of direction.

As a leader, you must put the fire department's interests before your personal preferences and act accordingly. What sits at the core of any successful organization is its principles and values, which are essential in guiding our priorities. Authentic leadership (see chapter 1) calls for leaders and stakeholders to work together with the employees to achieve mutual goals.

To rally a team around a common cause, you must have a clear vision and convey it to your team—constantly and creatively. Most people fail at conveying the vision constantly, considering that once they have delivered it, it is enough. Such people do not realize that not everyone in the organization shares the same level of familiarity with or passion for their vision. Most leaders are constantly on the lookout for the next challenge or opportunity, thinking that once they have communicated something, it has caught everyone's attention and been understood.

Ownership, Participation, and Employee Engagement

By prioritizing their employees' career advancement prospects, leaders exhibit commitment to the welfare of their team. Another critical aspect of leadership responsibility is supporting your workers by technically polishing their skills.

To bring out the best in your teammates and to encourage them to take up more responsibilities, you need to comprehend their objectives and promote their drive and innovative potential.

Empowering people who are following your directions is not impossible. In fact, leaders are in a position where they can aid their team members' professional development. Professional development is a means toward employee promotion. However, as a leader, you must demonstrate your dedication to expanding your team's reach by emphasizing advancement opportunities. It is also vital to put yourself and your team to the test on a regular basis.

When you collaborate with another, you are both more equipped to overcome your weaknesses since you are working together. A well-established professional development program will reward engagement, foster enthusiasm, and inspire continued advancement.

As a leader, it is your responsibility to bring out the best in your staff. Employee empowerment entails talking about personal and organizational ownership and putting it into practice in your everyday activities. To attain 100% results, you must provide 100% clear and correct directions. When assigning a task to your team members, you must be fluent in your communication as a leader. Furthermore, you must ensure that everyone participates and that no employee feels excluded from the organization. In sum, there are numerous strategies to involve the entire team while directing them to achieve success for themselves and the organization.

Eliminate freelance mentality thinking by forming *cross-functional teams*. These are groups that span departmental lines to fully utilize all your employees' ideas and talents. When assigning employees to these teams, encourage and reward them for taking on formal and informal leadership positions. This will also result in better communication within your organization, a higher ability to capitalize on opportunities, and the development of practical solutions to complex situations.

Nowadays, the most successful fire departments encourage all employees to take on leadership responsibilities. Not only will this relieve some of the pressure from the leader's shoulders, but it will also make your employees happier, more efficient, and more engaged with your organization.

Personal and Professional Accountability

Responsible and accountable leadership in the fire service needs to recognize the social contract they have with their communities. In many communities that support paid fire departments, firefighters often make well above the median income.

In other words, taxpayers continue to fund the expense of emergency response in these communities because they appreciate the service's intrinsic value. We are the protectors and guardians of that sacred trust and responsibility. Accountability with our s is a must and a cornerstone of community trust.

Our fiduciary responsibilities include forecasting what is yet on the horizon. Our communities are confronting comparable difficulties—such as the consequences of climate change, the housing crisis, the opioid epidemic, homelessness, and revenue loss—and the fire department is ideally positioned at the forefront of the response to many of these concerns. Moreover, each of the aforementioned difficulties provides the fire department with a unique opportunity to address the impact on the community. Early and deliberate interaction with community residents and other stakeholders, as well as governing bodies (e.g., city council), is essential for creating meaningful partnerships that value the fire department at the heart of the community. When budget constraints and allowances are considered, understanding community needs and demonstrating returns on departmental investments will pay off.

How much of a budget cut can we tolerate before compromising our service to the community and the safety of our organization's membership? This point is at the core of all budgetary debates. Conversely, what would our fire department look like if the sky were the limit? Since the chances of that happening are little to none, we must critically examine what we require to deliver service efficiently and safely, at the level that the community demands. Fight for what you want right now or fight later for what you don't want.

Budget demands that are not based on data-driven, evidence-based debate will be ignored. "The city will burn, and people will die" is a well-worn theme but has long outlived its usefulness as a talking point. People are looking for more than a passionate plea.

If we look at them through the correct lens, budgetary constraints and cuts become opportunities in adversity. If you're lucky enough to be part of the International Association of Fire Fighters (IAFF), their union leadership can be a terrific ally. The IAFF facilitates honest and open discussion between administrators and union leadership. Furthermore, NFPA standards should be used as best practices and desired (i.e., defendable) objectives for the community and organization. Respectful discussion should be the order of the day. Constantly defending your budget is a torturous loop, but it shouldn't feel like you're engaged in hostage negotiations.

Nothing is more powerful for your future than being a gatherer of good ideas and information. That's called doing your homework.

—Jim Rohn

In budget talks and negotiations, we should be able to responsibly—and reasonably—navigate questions that will arise. Specific efforts need to be made to avoid dominating the discussion, presenting condescending information, or attempting to relitigate the past. Do not try to impose the fire service's interest as the community's interest.

Professional growth requires us to stretch what we can accomplish as individuals and together. Doing this in a controlled environment is paramount (and allows failing forward, as will be discussed in the next section). This can occur on the training ground or in the fire station, through emergency simulations and tabletop exercises. Experiment: see what works and what doesn't, and develop adaptive solutions to nonconforming problems. The modern fire service doesn't need more rule books and manuals. Instead, the profession needs mission-focused and values-driven leaders at all organizational levels who are willing to experience failure. This requires the department and organizational leadership to embrace the ideology of *leaders' intent*—that is, empowering our firefighters and officers to make decisions supporting the purpose of the action, even if failure was the outcome.

Why You Should Seek Failure

More than want, we *need* people to be courageous enough to take calculated risks without fear of intense scrutiny and ridicule. Firefighting is an action that describes a profession. We need individuals to act. Fear of failure should not impede movement. Nor can the health and well-being of each firefighter. And that's why you—and your team—should seek failure.

Contrary to popular belief, success is not always the inverse of failure! True success can mean different things to different people, depending on their goals and approach to life. Now you may wonder, what is the value of success then? How can we evaluate success if it isn't measured against failure? That is, the bigger the success is, the greater will be the risk of failure and its consequences.

The fire department finds it difficult to accept that failure has value. There's a lot of pressure in the ranks to get it right. Therefore, providing our firefighters and officers with the opportunity for and reward of failure may at first seem at odds with the fire service's culture. So, why should we go out of our way to fail?

In answer, while it may be easier to discuss success and to envision this for our departments, failure must also be celebrated to appreciate successful outcomes. Success that is repeated and unchallenged can make you muscle bound, while failure creates agility in mindset and action. The importance of failure varies depending on the task.

Fast Failure versus Intelligent Failure

How can the fire service learn to embrace failure? In the business lexicon, the terms *fast failure* and *intelligent failure* are mechanisms to embrace failure purposefully and grow from it. These two practices teach individuals to see failure as an intentional consequence of experience and a knowledge-building process: The intent of fast failure is to cultivate an instinctive reaction that is in contrast to the traditional responses to failure of fear, shame, blame, and dysfunction; intelligent failure incorporates a "fail early, fail often" mindset, and organizations that embrace this can fill gaps in learning innovation and create change agents that can fail intelligently, providing long-term benefits to the team.

Another way to look at failure is through the lens of *failing forward*, which derives from Maxwell's book, *Failing Forward: Turning Mistakes into the Stepping Stones for Success*. Maxwell motivates his readers to move beyond the idea that failure is a negative and instead consider that failures are chances to learn. He wants people to see that failure is a positive way for people to learn, grow, and continue forward until they succeed. This is similar to the saying "fall seven times, get up eight."[2] Whereas Maxwell's framework entails changing one's mindset, Heath suggests that the best way to allow a culture of failure is to cultivate one where people feel comfortable acknowledging errors; then they can work to eliminate these in the future by learning and growing.[3]

The more we change our mindset to excite ourselves about the prospects of failure and its teachings, the more we will grow. We should all see failure as a positive force in the world around us, even in the fire service. Making a mistake or learning about a better methodology is simply a matter of growth. Regardless of the failure, we should all be open to learning how to improve. This comes from Maxwell's growth mindset—in other words, the choice to stretch ourselves beyond what we are currently capable of in pursuit of an extraordinary path for our potential development.[4]

Growth as a professional and a person requires us to stretch what we can accomplish—to push our growing edges. Cultivate a growth mindset and adapt this as the only way that we view our professional and personal development, even in the face of failures. Embrace the idea that we are works in progress, not finalized, as the leaders and team members that we are today.

As a chief officer, I expect those that serve their communities to take action, regardless of rank. What they need and should expect is support. Not for when things go according to plan, but rather for when we experience a failure. In fact, failure should be another reason to try, to work harder, and to test ourselves in controlled spaces so that we can achieve greatness and serve the public to the best of our abilities. We should train and develop assertive, values-driven, mission-focused fire station officers. Along with clear expectations and the support of the leader's intent, this will make it much easier to pull in on the reins than push timid individuals forward.

Seeking failure and embracing innovative responses to unanticipated or undesired outcomes offers the intangible benefit of maximized learning from failed outcomes as well as building personal and organizational tolerance and resilience to risk-taking. The aphorism "Don't practice until you get it right; practice until you can't get it wrong" speaks directly to the resilient mindset we as firefighters must embrace. Simply replace "practice" with "fail" in the previous quote, add a picture from the training ground, and we could have a motivational poster that should be in every fire academy. Every failure was a practice attempt to get it right. Every failure was another chance to perfect our skills, to learn, and to develop into even greater professionals in our field.

Whether in the fire academy or in the field, we need to train to the point of failure. If we want quantitative results, then we need to evaluate qualitative skills. As a profession, we need to make allowance for failures; if we don't, we create an environment that promotes inaction. Our collective success is on the line, as professionals and, more broadly, as a profession. Being successful motivates and inspires others; nonetheless, through our failures, we can relate to each other.

On the one hand, high risk should be associated with a high cost of failure. On the other hand, low-risk actions or decisions should be related to a low cost of failure and a high opportunity and value for lessons learned. Seeking failure should be advantageous. Increased perseverance, passion, grit, and determination are among the most important rewards. Learning through failure develops confidence in skill sets that can be applied in high-risk situations.

To be sure, the fire service expects and rewards successful outcomes, and rightly so. Positive—successful—outcomes are expected of public servants and professionals charged with public safety. Too often, the opportunities presented by failures are brushed aside or buried beneath layers of arrogance and ego. Selfishly failing to expose one's own failure sets others up to fail, robbing them of the gift of your failure as an opportunity to learn, grow, adapt, and move forward with confidence. Accepting and valuing failure builds a more resilient organizational culture.

Failure Is Just Another Opportunity to Learn!

As a leader, you must maintain tolerance for your employees' mistakes, as a necessary part of their learning and development. It is unnatural for an employer to believe that their employees will not make any mistakes. It's understandable that tests and interviews are used in the hiring process to determine whether the candidate is qualified for the job. However, the working environment will be new to that person at first, and it will take time to learn how things are done in this fire station. During this time shortly after being hired, the employee will make some errors. As a result, the employer must accept the worker's mistakes and give them time to adjust to the new environment.

As Maxwell shares, embracing failure distinguishes the average from the exceptional: "The difference between average people and achieving people is their perception of and response to failure." There is no fundamental mistake other than to not try at all, so when it comes to your development—or even your team's—create a culture where failure is accepted and where learning through one's failure is the goal to succeed in serving and supporting the public.[5]

Failure itself is a valuable opportunity to learn. We all need to learn to grow; otherwise, when we become leaders, we punish people for their failures instead of encouraging them to learn and grow from these experiences. This is an abdication of authentic leadership. Even if we work for an average leader who punishes people when they fail, that does *not* relieve us of the responsibility to learn from the experience. We should not overlook the critical need to learn why we willingly accepted a job working for average leaders who don't encourage people to learn from setbacks. However, as we will explore in the next chapter, neither should we overlook the need to prioritize firefighter health.

Chapter 7 Review Questions

1. Which of the following is the responsibility of an organizational leader?
    a. To prioritize personal preferences over organizational goals
    b. To lead without understanding the needs of their team
    c. To build relationships and create trust
    d. To avoid aligning with subordinates

2. How should a leader build trust and fidelity with an organization's membership?
    a. Ignoring the outcomes of decisions
    b. Avoiding the accountability process
    c. Consistently demonstrating accountability
    d. Not documenting any results

3. Which of the following is a necessary step to effect change in the accountability process?
    a. Not asking for feedback
    b. Praise only when outcomes are unfavorable
    c. Avoiding the repetition of the process
    d. Documenting both successful and failed attempts
    e. Maintaining a closed decision-making process

4. Which of the following is the primary focus of the Accountability "P" continuum?
    a. Placing blame on others
    b. Avoiding responsibility for decisions
    c. Taking ownership of outcomes
    d. Constantly changing the organizational values

5. What is organizational freelancing?
    a. A system where employees can freelancers in different organizations
    b. Encouraging employees to work independently without supervision
    c. Irresponsible actions within a permissive organizational environment
    d. Supporting employees to take on new challenges

6. Authentic leadership involves avoiding learning from failures.
    a. True
    b. False

7. Fear of failure should be an impediment to action.
    a. True
    b. False

8. Leaders should avoid forming cross-functional teams to utilize all employees' ideas and talents.
    a. True
    b. False

9. Constructive feedback should be vague and open to interpretation to encourage employee self-reflection.
    a. True
    b. False

10. Leaders who invest in holding employees accountable for their actions and providing feedback show that their actions and behaviors impact the organization's mission.
    a. True
    b. False

11. How can forming cross-functional teams benefit an organization in terms of creativity and problem-solving?

12. How does the aphorism "Don't practice until you get it right; practice until you can't get it wrong" resonate with the mindset of firefighters regarding failure?

13. How can embracing failure in a controlled environment contribute to developing adaptive solutions in the fire service?

14. How can accepting and valuing failure foster a more resilient organizational culture in the fire service?

15. How can organizations prevent normalization of deviation and promote a culture of accountability?

8 THE IMPORTANCE OF FIREFIGHTER HEALTH

Unlike many other jobs, firefighting is work in a demanding environment. Not only must firefighters be in exceptional physical condition to handle and avoid the hazards that they encounter on the job, but they must also have the spiritual and mental health to continue their pathway to success. Without a strong mentality, firefighters will struggle psychologically and may fail to prioritize their overall well-being. When this happens, firefighters do not deliver the results they need in their communities.

Over more than 25 years within the industry, I have seen how firefighter health affects team members and the department. Our priorities are imbalanced when the importance of firefighter health is overlooked. In my opinion, health is one of the most important priorities that each firefighter must commit to.

As mentioned, firefighters must maintain exceptional physical, spiritual, and mental health. Let's dive into these areas and explore what each entails.

The Physical Health of the Firefighter

The physical health of firefighters is important because it determines their ability to support the community. If not in top physical condition, firefighters would struggle to support their community, finding difficulty while facing the challenges that can be encountered on the job.

Firefighters have to not only exercise regularly but also eat healthily. Firefighters must ensure that they are fueling, resting, and caring for their bodies as much as possible. As a leader in your department, you have a duty both to yourself, to ensure that you perform at the top of your game, and to others, to ensure that you lead them appropriately.

For example, consider a fire chief training a new class of recruits. What aspects of health and wellness should the chief brief them on as expectations for them on and off the job? Two answers likely comes easily to you: nutrition and physical fitness; as we'll discuss, injury prevention is a third, not to be forgotten. Let's explore nutrition first.

The Role of Nutrition in Firefighter Health

You are what you eat, so by that token, you should take proper care of your body through what you put into it. Food is fuel, and the kind of food you feed yourself will determine the amount and type of work you can accomplish.

As a leader, prioritize having healthier options around your department. Get rid of the junk food that usually floats in vending machines, tempting you and others. Eat more nutritious meals around your team and encourage them to bring in meals that will fuel them for the job they have to perform.

One of the easiest ways to ensure that your team is eating healthier is to educate them on it. Contrary to popular belief, not every firefighter will know what they must do to keep themselves healthy. Consider bringing in someone to speak to your team members about the importance of healthy foods and balanced meals. The more your team members are informed about the need to eat healthier and how to do so, the healthier their food choices will be.

Firefighters Must Be in Peak Condition

It's also essential to ensure that firefighters are in top condition. This is because the work of a firefighter requires peak stamina and strength, unlike other careers and industries. While this may seem implied, it must be addressed. No firefighter will be prepared to take on the responsibility of preserving and protecting the community if they lack physical health. Period. Our challenges are far too strenuous for us to be lackluster about how we train ourselves.

From the moment we begin firefighter training to our last shifts in the service before retirement, our physical health, stamina, and strength ensure that we are capable of the task. Firefighting is not for the faint of heart or for those who may see strength and stamina as less of a priority.

If you want to see your team members thriving, you must ensure that everyone is in peak physical condition. The easiest way to ensure that this is the case is to make the environment suitable for this expectation: Have weight equipment and exercise available to them at the department or find a way to access a local gym. Create contests and encourage friendly competition among team members, to push each other to do well and excel in strength and stamina. Make it lighthearted while impressing them that this is important to their jobs in the fire service.

The reality is that if your fire service team cannot be strong and exhibit stamina when it matters most, it will cost your department the opportunity to save a life. For that reason, create a culture where your firefighters prioritize excellent physical health.

Injury Prevention and Safety

Another important aspect of physical health that is often overlooked is injury prevention. Unfortunately, the nature of the fire service makes your team more at risk of certain injuries. Any number of injuries—physical and psychological—can happen. No amount of preparation can prevent every potential issue, but that doesn't mean you should not still prepare for foreseeable risks.

Functional fitness training is intended to prepare your team to mitigate situations where an injury could happen. That is why physical exercise for both strength and stamina remains crucial to every firefighter in the field: it can prevent the most serious of injuries or ensure that people at least reduce the severity of the damage they acquire on the job.

Injury prevention can be implemented in several different ways in your department. First, education is of the utmost importance. Education should not be limited to just your initial training but be incorporation within the activities of your team long after your initial recruitment and entrance into the department. Remind your team members of best practices and regularly improve the rules to prevent even more injuries.

Second, focus on having the right resources and support available to ensure your team can prevent injuries. Without the right resources, your team cannot prevent injuries. Ensure that they are accessible within your department and have them on hand when you are on the job and in an emergency response. Lead by example and use these resources yourself.

The more time and effort you spend on education and resources, the more prepared you—and your entire team—will be to prevent injuries on the job. However, it's also crucial to consider mental health, which is the foundation of your resilience as a team and in the face of great crises that you must handle for your community.

Mental Health: The Foundation of Resilience

I am not a psychologist, but I am a professional firefighter. In the field, there is a growing body of literature on firefighter mental health. The importance of mental health cannot be overstated. Without the maintenance of your mental health, you will struggle under the weight of the high-stress environment.[1]

Firefighters are thrust into high-stress environments every single day. Feelings of danger or uncertainty are not abnormal to them, but they can impact one's mental health in ways that are not visible to others—or even ourselves. A diabolical situation emerges if the team fails to support each other. Without proper support from their peers, supervisors, and others, firefighters who are already struggling with their mental health may be unable to speak up. Such internalization in turn contributes to a downward spiral in their overall mental health.[2]

Why Support, Inclusion, and Trust Are Vital

Firefighters need more support than has historically be provided to maintain their mental health and continue in their positions without further damaging their outlook on the profession. Ensuring psychological safety entails creating a supportive, inclusive, and trusting work environment founded on effective communication, training, and leadership.[3]

Prioritize active listening and clear, concise instructions. Your team members need to be heard, and you need to be clear with them about what you are seeing and what needs to be adjusted. Furthermore, make sure that your criticisms are constructive. You need to foster a sense of belonging and ensure that your team feels like a proper team. If you do not prioritize cohesiveness within your team, you'll see a lack of trust—but more important, you will also see dwindling mental health from even your finest firefighters in the service.[4]

As the leader, demonstrate what you expect from service members. Create a culture where your attitudes and behavior exude your values, mirroring your desire for a psychologically healthy, supportive environment. You can speak to your team as much as you desire, but if they see you going back on your word or acting out in ways that do not reflect your inspirational words of wisdom, do you think they will follow in your footsteps? As their leader, you must demonstrate what you would like them to emulate—from effective communication to active listening to inspirational, uplifting guidance. What you lead with is what your team will practice.[5]

Crew Resource Management and Mental Health

How do you identify existing and potential threats and then communicate them effectively to avoid or mitigate them? That is what *crew resource management* (CRM) in the fire service is all about. CRM is also imperative for the benefit of the mental health of all your team members.[6]

CRM is used not only by you, as leader, but also by all crew members. Everyone has a responsibility to communicate regarding the challenges they anticipate, as well as any plans they have to mitigate or avoid these potential issues. When CRM is implemented correctly, you will see a positive shift in employee morale and enhanced efficiency in your operations.

CRM is the foundation for psychological safety, not its end goal. Therefore, CRM should be an integral aspect of your fire service department. However, the measures required to protect the psychological safety of your team extend far beyond this concept.[7]

The genesis of CRM and how it was implemented in the airline industry reflect that this was a tool for survival. However, the evolved state of psychological safety goes beyond survival and is more about thriving. CRM can help you to reach the foundational level of psychological protection you desire, but it should be expanded upon greatly. Implement more methodologies that create an efficient, safe psychological environment beyond CRM, and you'll see an increase in the number of your team members who are thriving.

How to Create a Psychologically Safe Work Environment

Beyond CRM, other factors are essential for a psychologically safe work environment. Let's examine these and how they can promote a better work environment for all fire service members.

The Presence of Strong Leadership

As detailed in chapter 7, leadership is perhaps one of the most critical factors in maintaining a psychologically safe environment for all your team members. Strong leadership sets an example. In additional, keep an approachable management team for your team members. When employees do not feel safe enough to come forward or think that their work environment is poor and unsupportive, this is more likely associated with the example you set through your leadership, rather than a reflection on them.[8]

Encourage your team to approach you—rather than have them sit and wonder how they can overcome their challenges alone. To mitigate stress, support your team in all high-stress situations and be transparent with them when you have overcome high-stress problems. Your honesty, leadership, and courage will all impact your team members and allow you to lead them toward psychological wellness within a robust and supportive environment.

Suppress the Stigmas

One of the most common issues that inhibits the creation of psychologically safe environments is the continuation of stigmas that make mental health issues unacceptable to talk about. The last thing you desire for your team is for them to feel that their mental health issues are a sign of weakness. If these stigmas are allowed to persist, people are more likely to struggle on their own, stay silent, and be unable to overcome their challenges. That's why you need to suppress the stigmas.[9]

You must lead your team to disengage from these stigmas. Dismantle these harmful ideas in your work environment by being open about the challenges of the fire service. No other profession has many of the same work hazards and challenges as the fire service, so you must address these concerns. Do not leave your team members feeling as though their problems and their challenges are theirs alone. By addressing the stigmas, you let your team know that their psychological safety matters and you reinforce that opportunities to receive help and support are present within your fire department.

Provide the Right Support and Resources

In the fire service, another imperative to ensuring that you have a psychologically safe environment, beyond addressing and squashing the stigmas, is to ensure that the right resources are present. You want your team members to be able to handle their concerns, manage their stress, and maintain their overall mental health. To do that, you need to make available a few specific resources and avenues of support.

Whether in-house or by referral, provide access to counseling services, support groups, and mental health training. The importance of these resources cannot be overstated. Without them, many team members may be unable to understand and cope with their stress and trauma.

In additional, you can also have in-house initiatives that do not require outside professional contacts. Hold debriefing sessions in the field and after high-stress occasions. Make this a regular part of your shifts to ensure that your firefighters have a chance to speak about what they experienced and come to terms with any stress or anxiety they may be experiencing so that they can reduce this as well.[10]

Jada Hudson's book *Firefighter Emotional Wellness: How to Reconnect with Yourself and Others* provides an in-depth guide to maintaining and improving the emotional well-being of firefighters.[11] Hudson, a licensed clinical professional counselor and certified alcohol and drug counselor, focuses on the unique emotional challenges faced by firefighters, including posttraumatic stress disorder (PTSD), and offers practical strategies to address these issues.

Jada Hudson's Principles of Firefighter Emotional Wellness

- *Understanding emotional impact.* Hudson emphasizes the importance of recognizing the profound emotional effects of firefighting. Firefighters often face traumatic situations that can lead to stress, anxiety, and PTSD. Understanding these impacts is the first step toward addressing them.
- *Destigmatizing mental health.* A crucial aspect of Hudson's approach is removing the stigma associated with discussing mental health issues. By sharing personal stories and professional experiences, she encourages firefighters to openly talk about depression, anxiety, addiction, and suicidal thoughts without fear of judgment.
- *Building resilience.* There is a difference between coping and resilience: While coping involves managing stress in the moment, resilience is about developing long-term strategies to bounce back from adversity. Hudson provides techniques for building resilience, such as mindfulness, yoga, and accelerated resolution therapy.

- *Practical self-care strategies.* Hudson outlines practical self-care strategies tailored to firefighters. These include proper nutrition, regular physical exercise, and mental health check-ins. She also discusses the importance of hobbies and activities that provide relaxation and joy outside of work.
- *Support systems and relationships.* Strong support systems are important both within the fire service and in our personal lives. Hudson advises on how to maintain healthy relationships, the role of peer support, and the importance of family involvement in a firefighter's emotional well-being.
- *Substance abuse and addiction.* Addressing substance abuse is a significant component of firefighter wellness. Hudson discusses the risks of addiction and provides guidance on recognizing signs of substance abuse and seeking help.
- *Suicide prevention.* Hudson provides critical insights into suicide prevention, offering strategies for identifying at-risk individuals and interventions that can save lives. She emphasizes the need for a supportive and proactive approach within fire departments.
- *Planning for retirement.* The transition to retirement can be challenging for many firefighters. Hudson discusses late-onset stress symptomatology and provides advice on planning for a fulfilling and emotionally healthy retirement.

Hudson's book serves as a comprehensive resource for firefighters, their families, and mental health professionals working with first responders. By focusing on these key areas, she has enhanced the overall emotional wellness of those serving.

Fostering the Right Culture

A psychologically safe environment is not possible without the right culture. We've talked about leadership, support, and more, but you also need to ensure that your organization is being guided toward the right culture for your team.[12]

What qualities foster a strong, psychologically safe environment? First, value and reward openness and transparency. Both the most experienced and the newest team member should be comfortable with sharing and discussing difficult topics with others. Furthermore, ensure that people feel encouraged, valued, and heard. This is achievable through open communication, which fosters greater trust and camaraderie among your team members. These steps put your organization on the way to the right culture, and your team members will feel safer psychologically.

Addressing Conflicts and Negative Behavior

Perhaps one of the most overlooked concerns that fire service leaders must handle is the mitigation of conflicts in the workplace. With all the psychologically challenging situations that firefighters end up in throughout their careers, the last place that fire service members should feel uncomfortable or unsupported is in the organizational culture. One of the ways that fire department culture can become tainted over time is by avoiding dealing with conflicts.

When conflict arises or any negative behavior occurs in your culture, address it promptly. Do not wait to tackle such issues. Let everyone in your agency know that they are appreciated and valued. Also maintain expectations with confidence in your culture to foster change and psychological health.[13]

Addressing conflict in the fire service is crucial for maintaining a functional and cohesive team. The following two strategies are effective for handling conflict within our organizational environment:

- Open communication and mediation
- Training and development

Open Communication and Mediation

Open communication

- *Encourage direct dialogue.* Encouraging open, honest, and respectful communication is essential. Team members should feel comfortable expressing their concerns directly to each other. This can prevent misunderstandings and build trust.
- *Regular meetings.* Hold regular meetings where team members can voice their concerns and discuss issues in a structured environment. This can include debriefings after incidents, which provide an opportunity to address any conflicts or other concerns that arose during operations.
- *Feedback mechanisms.* Implementing structured feedback mechanisms, such as anonymous suggestion boxes or regular check-ins with supervisors, can uncover conflicts early and allow them to be addressed promptly.

Mediation

- *Third-party mediators.* In some cases, bringing in a neutral third party to mediate can help resolve conflicts. This person can be a trained

mediator or someone from another department who has no stake in the outcome.
- *Structured mediation processes.* Establishing a formal mediation process within the department can ensure that conflicts are handled consistently and fairly. This process should be well-documented and communicated to all team members.

Training and Development

Conflict resolution training

- *Workshops and seminars.* Providing regular training sessions on conflict resolution can equip team members with the skills they need to handle disputes effectively. These sessions can cover topics such as active listening, negotiation techniques, and emotional intelligence.
- *Role-playing scenarios.* Role-playing exercises during training can help firefighters practice conflict resolution skills in a safe environment. This prepares them for real-life situations and enhances their ability to manage conflicts on the job.

Leadership development

- *Leadership training.* Training officers and senior firefighters in leadership and conflict resolution is critical. Effective leaders can often prevent conflicts from escalating by addressing issues early and modeling positive behavior.
- *Mentorship programs.* Establishing mentorship programs can help new or less experienced firefighters develop strong communication and interpersonal skills. Mentors can provide guidance on navigating conflicts and fostering a supportive team environment.

Addressing conflict in the fire service requires a proactive approach, combining open communication, mediation, and comprehensive training programs. By fostering an environment where team members feel heard and supported and by equipping them with the necessary skills to handle disputes, fire departments can maintain a strong, cohesive, and effective team.

With that in mind, also reward positive behavior. In addition to rewarding team members for their efforts, reward them when they speak up. This can further the change you wish to see in your culture and allow your team members to enjoy better mental health in the long term.

Continuous Evaluation and Improvement

Last but certainly not least, always be focused on continuous improvement. Your culture will shift over time, so you must constantly identify how and what needs to be done to continue supporting your team. Two of the best resources that have assisted me in this effort are *The Fearless Organization*, by Dr. Amy Edmundson, and *Turn the Ship Around: A True Story of Turning Followers into Leaders*, by Captain David Marque of the U.S. Navy. Both offer great insights into leadership and the importance of psychological safety[14]

A Final Note on Promoting Psychological Safety

Promoting psychological safety might be more difficult than you imagine, yet it is a critical aspect of the firefighting service. You must ensure that your firefighters have great well-being and can manage stress while being safe. By adhering to the tips in this section, you can instill the right foundation to foster a psychologically safe environment. Admittedly, it's not simple, but it certainly is worthwhile for the future longevity and cohesion of your entire team.[15]

Moral Injuries in the Fire Service

While not a new term, *moral injury* is one that is not commonly discussed among many scholars in the field. This is best defined as the psychological stress that occurs when an individual is exposed to an event that violates their moral or ethical beliefs. It's not PTSD, but it can coexist with this disorder.[16]

It would be a mistake not to deal with moral injuries when discussing psychological safety. In the fire service, we are trained to protect life and property from fire hazards and other emergencies, but we are constantly exposed to traumatic incidents. These impact us immediately, in the moment, and they can have effects that last for decades after.

As firefighters, we struggle with many of the same challenges as do *wounded healers*, as described by Jung. Wounded healers are driven by the suffering of others to help them while also experiencing grief in their own lives. Our deep empathy for others and how we intervene when the emergency strikes is often rooted in our own adverse experiences.[17] We face the physical and emotional wounds of our demanding profession willingly, driven to heal and help others while also demonstrating remarkable resilience and compassion. The moral injuries wounded healers endure can occur within any traumatic situation, from child abuse to domestic violence to fatal accidents.

The Causes of Moral Injuries in the Fire Service

Moral injuries can happen in several experiences that firefighters encounter daily on the job. Loss of life, property destruction, and injuries to ourselves are only a few such instances where we might sustain moral damage. While moral injury is a new concept in the fire service, those incurring moral injuries may have a difficult time adjusting to civilian life and develop symptoms related to depression and anxiety. Furthermore, they may also have an increased risk of suicide.[18]

Moral injuries can occur even in minor encounters, such as witnessing unethical actions by peers or superiors. While we are expected to act with integrity and ethics, there are times when firefighters fail to do so, leading to moral injuries. Many feel betrayed, disillusioned, or even angry at their superiors. Other times, fire service members may also endure moral damage when encountering situations challenging (e.g., directly ignoring or negating) their morals or ethics. Examples include harassment and discrimination.

The most unfortunate situation leading to moral injury is when firefighters have attempted to save someone's life but been unable to. The difficulty of allocating resources and acting during the crisis is often to blame, but firefighters still feel guilty or shameful about their role (and what they did or did not do). In these crises, support and resources are of utmost importance to their mental health; team members should be encouraged to speak about what they have experienced and how they feel after these traumatic experiences.

One of the most preventable moral injuries is also one of the most likely, happening within the fire department culture. Remember what I said about ensuring that your culture is strong, supportive, and indicative of outstanding leadership? Without the proper guidance, firefighters may find themselves in situations where they witness unethical behaviors, poor supportive culture, and activities about which they feel they must be stoic, even when they need to speak out. Part of this is the "good firefighter" stereotype, which leads team members to feel like they must live up to an unattainable image. Feeling the pressure to conform can not only disenchant your fire team members but also lead to poor psychological safety. You must ensure that your culture is not the culprit for your team's inability to enjoy a psychologically safe environment.

What Are the Impacts of Moral Injuries?

Moral injuries can do more than cause bad days for your team members. Firefighters with moral injuries suffer from poor mental health, anxiety, depression, PTSD, and other challenges. Moreover, this can also extend into other areas of their life, such as their long-term goals and careers.

Firefighters who experience moral injuries might feel less motivated, have less job satisfaction, and even present lower performance. This can be easy to spot in your team and is another indicator that they lack the support they need from your curated fire station culture. Moral injuries can also impact your firefighters' relationships with others, extending into how they interact with their coworkers, you, their friends, and their families.

With a moral injury, individuals tend to withdraw from their environment. They may stop being as social with others around them or may struggle to maintain their relationships. This can lead to isolation or detachment, which can cause more challenges. Many may even feel that they are without a sense of purpose in their lives. Some might even feel compelled to leave the fire service but may be unable to because of difficulty adjusting or the lack of structure in other professions. That's why it is crucial to find ways to prevent moral injuries and manage those that already exist.

How to Prevent Moral Injuries in the Fire Service

Moral injuries can be prevented in the fire service through various strategies that can be incorporated into your culture. First, one of the most important is to educate your team and train them on the methods for overcoming or addressing known moral injuries, such as loss of life or property. Allow them to understand the causes and impacts of these injuries and what remedies may be possible if they speak up about their challenges. Education is the quickest way to assist your team with coping with these challenges.

Creating a supportive environment comes next, as this will allow people to be included in the team effort while also speaking up without fear of retaliation for what is shared. You should instill clear policies to ensure that you can address these issues and prioritize your team's mental health. Connect to resources that would help your team while ensuring everyone has access to what they need. Such an example is through programs such as *employee assistance programs*. These programs can be tailored to your team as well!

Finally, implement evidence-based practices to treat and manage health conditions. Examples include counseling and peer support groups. Peer support groups can be fantastic ways to promote a more robust team culture. These groups are great for connecting with others while allowing your team members to feel heard, understood, and valued. You can also provide training for your peers on active listening, empathy, and even resilience. Through training and building resilience (see chapter 14), you can encourage your team to adopt strong coping strategies that help them emotionally regulate, prioritize self-care, and understand their sense of purpose in their work.

Cognitive behavioral therapy is another integral, peer-reviewed, and proven method that allows your fire team to discuss their challenges and focus on resolving their concerns. By connecting team members to counselors who practice this form of therapy, you can facilitate treatment of their moral injuries. This allows them to regain their positivity and motivation once more.

Aside from these interventions, remain focused on leadership support including regular debriefings (as mentioned above, in "Provide the right support and resources"). How you guide your team as a fire service leader indicates how your team will act. This also relates to ensuring your team members have a solid work-life balance. Firefighters should feel they have time to spend with family, pursue their hobbies, and care for their mental health.

As fire service leaders, you must focus on ongoing support as well. Mental health and wellness are much more than one-time considerations; they require leadership to commit to maintaining a holistically healthy workforce. Moral injuries will not just appear and disappear once but will be a continuous issue for as long as you are a part of the industry. It may be a sad reminder of the sacrifice we all make in the fire service, but it is also a chance to improve your support for your brothers and sisters in the fire service. However, many ways exist to promote psychological health and reduce moral injuries.

Spiritual Wellness in the Fire Service

Spirituality might not seem pertinent to the immediate health of your fire department. Still, it can be a great tool to encourage firefighters to find and commit to a purpose while feeling connected to your team effort. As a personal aspect of each person, spirituality is often seen as separate from the workspace, but I encourage you to explore how this can impact your organization.

Religion or spirituality is an aspect of our individual uniqueness. It is unlikely that everyone in the department will have the same faith or spiritual beliefs. Regardless, each firefighter's faith will impact them in similar, general capacities. With faith, individuals feel a sense of morality and can easily decipher right and wrong. Many also achieve a distinct sense of purpose through their work.

In your organization, you can leverage your firefighters' uniqueness and the power of their spirituality. Encourage them to have faith, practice their faith, and let their spirituality guide them. Although you may not engage in spirituality discussions regularly, when there is respect for each others' unique beliefs, they can be a means of connection. When times are tough, you can turn to your faith as a source of support. One of the most common ways you can support each other and your respective faiths is through prayers, reflection, or meditation. As a team, you can engage in these practices or, at the least, lead your team to see them as beneficial to the department.

Encourage your team members to lean into their spiritual practice as much as needed. Importantly, faith is inherently personal. Still, you can allow your organization to connect to what they believe and use this as a foundation for strength during their careers.

If you would like to refrain from discussions on religion as much as possible in your organization, consider focusing on self-care and balance. While not inherently spiritual, balance and self-care are needed by all firefighters. When paired with your emphasis on mental and physical health, you can encourage your team to remain emotionally healthy and satisfied in their careers and their lives beyond the fire service. Be the first to promote and support your fire team in engaging in activities that bring them joy and fulfillment—even beyond the four walls of your fire station. These small changes and intentions will bring astronomical benefits to your entire organization.

Curating a Positive, Healthy Culture

Embodying the ideas discussed in this chapter will allow you to advance to a new level as a leader. With emphasis on physical, mental, and emotional or spiritual health, you will see a positive change in your organization. Moving beyond the healthy culture you are cultivating, the next chapter will discuss the five laws of leadership in the fire service and the need for adherence to them.

Chapter 8 Review Questions

1. Why is mental health described as the foundation of resilience for firefighters?
   a. To prioritize physical well-being
   b. To enhance job performance
   c. To withstand high-stress environments
   d. To improve decision-making skills
   e. To establish hierarchy within the team

2. Why is crew resource management (CRM) considered a foundational aspect of psychological safety but not the end goal?
   a. CRM is not relevant to psychological safety.
   b. CRM focuses solely on physical safety.
   c. Psychological safety encompasses broader measures beyond CRM.

3. Why is education considered an important strategy for preventing moral injuries in the fire service?
   a. It does not help firefighters understand the causes of moral injuries.
   b. It weakens the team's ability to cope with challenges.
   c. It assists team members in understanding impacts and addressing challenges.
   d. It promotes a culture of fear and retaliation.
   e. It discourages seeking support from employee assistance programs.

4. What approach does Hudson advocate to destigmatize mental health issues among firefighters?
   a. Avoid discussing mental health issues openly.
   b. Encourage judgment and criticism toward mental health struggles.
   c. Normalize discussing depression, anxiety, and other issues.
   d. Punish firefighters for talking about their emotional challenges.

5. Which of the following is critical when dealing with moral injuries in the fire service?
   a. Downplaying the importance of moral injury
   b. Avoiding discussions about traumatic incidents
   c. Ensuring well-being and safe stress management

6. Creating a supportive environment within the fire service allows individuals to speak up without fear of retaliation.
   a. True
   b. False

7. Access to counseling services, support groups, and mental health training is not important in ensuring a psychologically safe work environment.
   a. True
   b. False

8. Peer support groups can be beneficial in promoting a more robust team culture and helping individuals feel heard and valued.
   a. True
   b. False

9. Strong leadership is not crucial in maintaining a psychologically safe work environment.
   a. True
   b. False

10. Implementing evidence-based practices such as counseling and cognitive behavioral therapy can help manage moral injuries in the fire service.
    a. True
    b. False

11. How does effective leadership foster a mentally healthy work environment for firefighters?

12. Why is it crucial to suppress stigmas around mental health in a fire service team to foster psychological safety?

13. In what ways can understanding emotional impact benefit firefighters in managing stress and anxiety?

14. In what ways can continuous evaluation and improvement benefit the overall effectiveness of a fire department?

15. How can moral injuries impact a firefighter's mental health and overall well-being?

9 FIVE LAWS OF LEADERSHIP

In the modern dynamics of today's corporate world, staying relevant and compliant is extremely important. *Compliance* means adhering to ever-changing standards, regulations, and laws. Note that compliance is a choice. Once individuals are deemed competent at the skill, behavior, or expectation, they remain compliant.

Compliance with desired behaviors must be tempered with practical knowledge, application, and skill, not obsequious or robotic behaviors. Compliance with organizational and societal rules and laws must be justifiable and practicable.

Organizations need mechanisms and leadership to recognize *noncompliance*, or deviation, and use judgment, discernment, and organizational values to determine an appropriate course of action. After all, noncompliance can be unavoidable in some circumstances, whereas incompetence should be recognized and rectified.

Ours is a profession based on mission-driven decision-making. It would be lazy and potentially deadly to always choose compliance over discernable outcomes. Therefore, organizations should trust and support leaders' intent. If noncompliance was a conscious decision, there might be an appropriate reason for the action.

Addressing Incompetence and Noncompliance

For supervisors, officers, and command staff, it is crucial to determine the root cause when you recognize an issue. Almost all disciplinary actions can be recognized as a lack of either competence or compliance.

If an employee or subordinate is incompetent, your organizational responsibility is to ensure that they become competent. Devote whatever time and methods—which may creative (as long as legal and ethical)—are necessary to create skilled employees. Consistently apply what consistently works. This is your burden and responsibility.

Once an employee is made competent, it is the prerogative of the employee to choose to be compliant or not. Discipline should be administered only when a qualified employee chooses not to comply with the organizational mission, vision, values, rules, and expectations.

You must provide clear, concise expectations to work toward competence if they are not competent. For example, as a basic competency, your organization may require firefighters to maintain their turnouts in a prescribed manner supported by NFPA 1581.[1] Once the firefighter understands the expectations and requirements and this knowledge turns to action, they have been made competent. If you discover that they have chosen noncompliance, they have chosen to deviate from the same expectations later. This distinction can be a great tool in determining who needs discipline and training.

Developing Competence and Choosing Compliance

Competence and compliance comprise knowledge, skill, talent, training, and limitations. Delivering tangible answers to the how, what, where, and why of our profession is essential to our organization and communities. Knowing the organizational and ethical limitations is critical to identify boundaries and allow executive leadership to hold the line.

For self-reflection, when encountering an issue, ask yourself, Am I qualified and competent, and am I compliant? If not, where was the breakdown? Is it a training, education, or competency issue? Or, for one reason or another, did you choose to be noncompliant?

Competence is developed over time. Organizations and individuals build their competencies through training, responding to incidents, continuing education, and qualification or promotional processes. In the early phases of our careers, we need direct supervision and immediate feedback. This need for oversight should taper off over time as we develop our competencies.

Ensuring that our membership complies is another matter. Rules and regulations must be considered compulsory and supported by organizational mission, vision, and values. Otherwise, they are just words on paper (or a screen). Compliance with corporate requirements means adhering to an organizational minimum standard.

Continuous training and experience build competence. Furthermore, as professionals, we must stay current on industry trends and knowledge. The membership seeks to develop their competencies, and supervisors and leaders hold them accountable for compliance. In other words, being competent and qualified in a profession are skills, whereas compliance is always a choice. When both competence and compliance are present, this produces tangible dividends for the organization's leadership and membership.

Taking Ownership Is Half the Leadership

Taking complete ownership of your outcomes by holding no one but yourself responsible for them is the most powerful thing you can do to drive your success.

—Gary W. Keller

Ownership implies pride in what you do, responsibility for your actions, and accountability. *Taking ownership* of your actions means holding yourself solely accountable for your mistakes—whether intentional or unintentional, consequential or inconsequential. Taking ownership of your work can manifest as proactively seeking viable solutions and creative approaches to your problems and challenges, planning for future issues, being attentive and responsive, and most important, being downright helpful. In addition, this practice works on other gray areas of your life. In general, when you stop holding others—whether people, objects, surroundings, or life events—accountable for the consequences of your decisions, you gain a sense of authority.

Former Navy SEALs Jocko Willink and Leif Babin introduced the idea of *extreme ownership*, a principle that states that leaders must take ownership of everything under their command. Personal responsibility—for successes and failures—is of utmost importance to inspire team members not only to succeed but to take ownership of their actions.[2]

One tenet of extreme ownership is decentralized control. Remember the discussions on accountability and leading by example? (If not, refer back to chapters 7 and 6, respectively.) Leaders instill an example, as extreme ownership takers who take pride in their efforts, and this leads others in their organization to follow suit. Eventually, you have a fire station of individuals all pridefully taking extreme ownership, yielding astronomical benefits.

Someone who takes responsibility and ownership does not expect anything to be given to them, nor do they take any step in life for granted. These individuals expect that it will take hard work to accomplish their goals, growth to become a better leader, and going outside their comfort zone to reach success. Ownership takers do not play office politics or find blame in finances, upbringing, or the availability of resources, among other excuses. If they are going to achieve their goals and realize their desires, they are going to be responsible for when, where, and how this happens.

Those who take ownership understand that they control their efforts and attitudes. Only they are to be held responsible for their actions and their reactions to those actions. Whether you aim for a promotion or seek a new job, taking ownership is a way to demonstrate your passion and work ethic to your bosses (current or future).

Your attitude and drive make you stand out among your peers and are attributes that enable your career growth. Admittedly, having the skills and knowledge base to survive in any workplace is imperative. However, taking ownership of your actions and doings requires no preexisting skills or particular knowledge, degrees, or certifications; rather, these can be learned and acquired over time. With a growth mindset, you'll see that most skills you desire to acquire are within reach with hard work and discipline. All that it requires is the *effort*—a lot of it!

One of the most crucial tasks of leadership development is influencing the leadership skills of team members to meet the organization's demands and yield the desired outcomes. Although motivational factors and processes may vary between organizations, depending on the culture and nature of the task, certain leadership qualities will be present in every successful leader. Revitalizing the spirit of teamwork and unified efforts is a much more complex task than it may sound. For an effective leader, their behavior must reflect on their conduct, influencing people to change.

When leading a group of individuals toward a main objective, certain rules apply to you. First, the goal of a leader is to understand and share vision with team members, by helping them to do all of the following:

- Align their vision to yours
- Take ownership of the vision
- Contribute to the vision
- Communicate the vision

Second, successful fire service organizations are built on a foundation of accountability, and accountability is not conferred by rank or title—not by a long shot. Taking responsibility for one's actions and for their consequences is a defining trait of accountability. It is not a "once-in-a-while thing," to be used only when convenient or practical. Consistent and calculated actions and efforts define accountable leadership, which will have an effect on accountability of team members. Employee engagement, participation, and ownership will be affected if they witness or experience a lack of accountability. Ultimately, it will be followers, not leaders, who bestow the credential of accountability.

Third, being committed to and responsible for your actions means being committed to and responsible for the organization's values and mission. The right people are assigned to the right tasks, thanks to organizational leadership. This structure sends a strong message about the importance of the job and, more important, the results.

Finally, being well intentioned or having good intentions is not the same as being accountable (for more on this distinction, see chapter 7). Personal and organizational accountability is a commitment to uphold the organization's mission and values.

A Leader Owns Responsibility and Outcomes

An efficient and attentive leader always needs to know where they are leading their employees. Therefore, you must clearly envision the fire department's direction and individual developments. Further, this vision must be compelling to all employees involved, not only those at the top. If employees do not buy into the vision that you have created, seek their insight and create a new one. Find a way to connect with them, to show them the power of your goals, and to encourage them to commit to the idea.

Additionally, a leader needs to get over the habit of selflessness in the workplace. What benefits any one member of the team will profit the entire workforce and the organization. This goes beyond the leaders themselves and stretches into the way that each employee and member of the team can advance the entire team.

The leader needs to communicate their vision to others—in other words, engaging in vision casting (as detailed in chapter 6). Moreover, as the market strategy alters, so does the fire station's strategic landscape. You must modify your vision as the organization evolves and share a brand-new future for the fire department to work toward.

To succeed, leaders might apply transformational leadership. Visionary leaders may also play a vital role in the continual effort to create a team vision that matches the organization's goals at every step of the way. After all, visionary leaders and those who subscribe to transformational leadership believe that the team and the goals constantly evolve. That also means taking critical accountability when the team progresses, adapts, and grows into the next best version of their capabilities and combined efforts.

A leader is to be held accountable for everything that occurs in a workplace. This means that if you or a member of your team happens to make a crucial mistake, as a leader, it is your responsibility to take the blame, instead of pointing fingers or allowing anyone to criticize other included members. Blaming other team members not only harms the person you are publicly accusing and humiliating in a working environment but also fosters an unhealthy atmosphere where employees feel scared to make a mistake. This is wrong in every sense: it takes away a person's right to feel comfortable in the workplace, and it prevents them from learning not to repeat the same mistakes. In other words, when leaders do not take ownership and instead cast blame, they destroy the nurturing culture that they worked to foster, and employees will not be compelled and comfortable coming forward when they have an issue. When the blame game is afoot, employees hide their feelings and undercommit, which is detrimental to the entire team and the service the fire department delivers to the community.

As a leader, you must present a culture of courage in the workplace. Nurture a workplace environment where people can speak up freely and share their opinions and concerns; employees who feel heard are far more productive and creative. Without a culture that values employee input and feedback and encourages growth and development, teams suffer—and by and large, so does the entire organization. When mistakes happen, have team members discuss what went wrong and devise a plan of action, rather than blame others. In a culture of courage, your employees will own their mistakes with pride and accountability. In short, exceptional leadership cultivates growth.

Extraordinary leaders are aware that they need to accept mistakes. They solve those mistakes without unnecessary drama. As a leader, you must learn to accept blame for the team and strive to bring remedial resolutions, even if you personally are not at fault. Regardless whether you messed up or a team member did, your approach to fixing the problem should be the same. If you are the leader, take responsibility for problems, for good and bad, because that is how you maintain an equilibrium in the workplace.

Leaders Must Adhere to the Organization's Values and Morals

Managers do things right. Leaders do the right things.

—anonymous

As a fire service leader, part of mastering your leadership skills is leading by example. This involves inspiring the people surrounding you to thrust themselves forward and strive for greatness. You must demonstrate the right way to them, blazing the trail for others to follow. When you lead by example, your team will undoubtedly follow you. Furthermore, when your team members are valued, they are loyal and in turn value you as a leader.

Being the leader, if an endeavor fails, you are rightfully held accountable and answerable. Once you have taken the lead, it is your duty, more than just a goal, to demonstrate your proficiency and expertise by giving directions and providing recommendations for the team members who will encounter predicaments in completing their tasks. Nonetheless, you should encourage the workforce to follow concrete steps that you mutually develop while intermittently checking back on their progress, considering extra guidance or a change in direction if required.

An honest leader holds an important part in an employee's life, inspiring them to greatness in life. Honesty is the ability to develop trust in followers and legitimacy. Honesty is interlinked with the concepts of positivity and integrity. Moreover, honesty in leadership entails transparency and openness. Honesty measures your willingness to communicate what you think or feel about something, even when it is uncomfortable, or the idea is considered unpopular. Another important aspect of honesty in leadership is the willingness to listen, share, and discuss ideas before the data are completely thought through. When available alternatives are not completely crystallized or when the discussions are not final, this is the time when your consideration toward your employees and how much their participation matters to you. Finally, honesty means keeping your word, following through on promises, and delivering on time—an integral part of any successful and influential leadership.

As a leader, you are always observed by your employees, managers, colleagues, subordinates, and others. You must lead honestly and by example so that you can expect the same in return from your team. If leaders are not honest about their ideas and vision, how can they be truthful toward their subordinates or others who consider them role models?

Chapter 9 Review Questions

1. How should a visionary leader modify their vision as the organization evolves?
   a. By keeping the vision stagnant to maintain consistency
   b. By sharing a brand-new future for the team to work toward
   c. By solely focusing on personal goals
   d. By avoiding any changes in the vision

2. Which of the following best describes ownership?
   a. Expecting things to be given
   b. Blaming surroundings
   c. Taking responsibility and pride in actions
   d. Avoiding hard work
   e. Depending on others for success

3. Which of the following is important for successful fire service organizations?
    a. Relying on rank or title for accountability
    b. Using accountability only when convenient
    c. Taking responsibility for actions and consequences
    d. Avoiding communication within the team

4. What is the relationship between honesty, trust, and leadership?
    a. Honesty is not important in leadership.
    b. Trust is not related to honesty.
    c. Honesty builds trust and legitimacy.
    d. Leadership is solely based on authority.

5. What should organizations use to determine an appropriate course of action in the face of noncompliance or deviation?
    a. Robotic behaviors
    b. Obsequiousness
    c. Judgment, discernment, and organizational values
    d. Blind obedience

6. Leaders need to inspire and set an example for their team by demonstrating the right way.
    a. True
    b. False

7. Continuously training and gaining experience will not significantly impact an individual's competence and compliance in their profession.
    a. True
    b. False

8. Leaders should not be held accountable when an endeavor fails, as it is a team effort.
    a. True
    b. False

9. Organizations should trust and support leaders' intent, even if noncompliance is a conscious decision with no justifiable reason.
    a. True
    b. False

10. Blaming other team members can foster a nurturing culture in the workplace.
    a. True
    b. False

11. How is honesty related to the concepts of positivity and integrity in leadership?

12. How can a leader inspire loyalty and commitment from team members?

13. What is the significance of compliance with organizational rules and regulations supported by mission and values in promoting a healthy work environment?

14. How can a leader balance accountability for both successes and failures within the team?

15. In what ways can a leader promote a culture of accountability and ownership within the team, especially when handling mistakes?

10 DEVELOPING YOUR TEAM LEADERSHIP ETHOS

It is not easy to be a great leader. When you have power, you also have responsibilities. Numerous matters must be managed, people supervised, and regulations followed consistently to achieve a positive outcome.

One characteristic that distinguishes a great leader is *counseling*. In the context of leadership, this entails focusing on the growth of your subordinates and educating them about how they can continuously improve in their field. This chapter concentrates on counseling and explains why it is a requirement for effective leadership. While mentoring and coaching subordinates remains beneficial, counseling them is different: When leaders counsel their subordinates, they are preparing future leaders to take their place when they retire.

One major issue that frequently perplexes leaders is the belief that this will create competition for them—and that their subordinates will replace them well before their retirement. This common misunderstanding prevents many leaders from entirely focusing on their subordinates because they don't want to lose their position in the fire department where they've worked for so long. However, leaders must realize that counseling others will only lighten their workload. By dividing the workload, everyone involved will have a much easier time. If the rest of your team members are strong enough and perform to their full potential, the fire department will not disproportionately rely on a small subset of the team during a critical period.

Importantly, a leader will offer advice to those who work for them only if they have empathy. In other words, they have to be sincerely concerned with the welfare of those being counseled. This also provides leverage to the person being counseled. This is because the leader constantly supervises at least one person and often more, but not all of them will be groomed by the leader through counseling. The others may grow through mentorship or training programs available to help them advance in the organization, but counseling is an entirely different matter, geared toward succession.

Almost every famous person in the world was mentored, trained, and counseled at some point to achieve success. Consider Arnold Schwarzenegger. He is well known around the world for being a great athlete, a world-class actor, and a politician. His story begins with leaving the Austrian military to pursue his dream of becoming a bodybuilder, which led to modeling opportunities, a career as an actor, and eventually, political fame as governor of California. But how did he manage it?

Schwarzenegger did not do it all by himself. Someone was always there to guide, mentor, and counsel him. Even though he is exceptionally talented, he was steered onto the best path by someone better than him. If you want to be a great leader or want your employees to be great leaders, mentorship is key.

Schwarzenegger's mentors or teachers had to change over time as the phases of his life changed. When he was an athlete and a bodybuilder, he had one teacher or mentor; when he decided to become a fashion model, his teacher or mentor would have changed, and then again when he decided to become an actor, as well as when he became a politician. All of these teachers and mentors were experts in their fields who helped Schwarzenegger prepare for a brighter future without hesitation; they all imparted valuable lessons that aided him in better shaping his future. It would have been impossible for one person to do it alone, but none of them could have stayed for his entire journey.

Many believe a person can only be a leader if they lead an entire team. However, the word "leader" means to lead in the right direction and it does not specify that a leader must be responsible for an entire fleet. Rather, they are recognized for their qualities and abilities at providing guidance.

Within any fire service organization, it is essential for successful management and leadership to motivate and empower their personnel. This can also improve work quality and productivity among employees.

Feeling motivated can help people to solve problems, create personal aspirations, and build expectations that align with the collective organizational objectives. For employees to work productively toward these objectives, leaders must have effective methods to motivate, encourage, and support. There is no one-size-fits-all approach, because everyone is motivated differently. That is part of the joy of leadership: you are constantly evolving to meet the needs and desires of your unique team while guiding them to success.

This assertion also applies to the community. In all emergencies, responders have the essential duty to provide urgent assistance, which can include serving as a chain of survival until victims are handed off to definitive medical care. A first responder's level of care, urgency, and willingness will be affected by their leaders' ability to motivate, support, and encourage their crews. For the fire service, boosting these attributes yields benefits for the greater community. To further explore this dynamic, let's next examine what leaders and managers can do to motivate their teams.

Dissecting Motivation

We'll begin this section by discussing what traditional methods of motivation exist for the target group of fire service teams. Later in this section, we will navigate the challenges and opportunities in motivating this community.

Motivation in the fire service won't be successful when a carrot-and-stick approach is used. Traditional leadership, as explored in chapter 1, will go only so far. You want to motivate your team, not create disconnects or stir up power struggles.

Leadership is about guiding teams away from punishable actions and toward desired behaviors and activities. That can work to modify the behavior of employees but not their motivation.

There is no nationwide or international standard for fire service training. Typically, each organization will incorporate some form of basic training. This is often extended to include a specialized skill set, such as fire behavior theory, basic EMT training, and related fireground operations.

Yet, the reality of a firefighter's role after leaving the training environment entails being first on the scene and entering an emotionally distressing environment, sometimes alone. Firefighters must maintain composure, gain control of the situation, and take action that initiates tactics for survival. Very few carrots and no sticks help navigate these situations. As compared with other industries, firefighters are not deciding whether to make a phone call to earn another sale; rather, they are fighting their instinct to run from fire instead of pursuing the danger to keep their community safe and protect its citizens. To motivate these exceptional employees, you must inspire them to act, to work hard, and to push past the fears that might hold them back.

The fire service, made up of various generations and backgrounds, is as diverse as the communities it serves. Similarly, the motivators of these individuals that sustain their ongoing commitment to the profession will also vary. Motivation is an individual experience; therefore, your efforts to motivate your employees must be tailored to their general needs, often even personalized to the individual.

A study conducted by Khalemsky and collaborators revealed that firefighters' motivations can be understood in terms of six psychological functions: values, understanding, enhancement, career, social, and protective.[1] In other words, first responders serve based on essential values and beliefs (e.g., altruism and humanitarianism); seek increased knowledge and the practice of unused skills, as well as psychological growth and development; gain career-related experience; and aid in social acceptance, develop social groups, and strengthen.

Altruism has been identified as a primary motivator in numerous studies. According to data from one study, giving back to the community, a desire to help, and community spirit are all prominent altruistic motivations. Other personal motivators include a desire to save lives, contribute to health care, and gain field experience in preparation for future health care–related careers. In your team, many of your firefighters will have been driven to join the force because of altruistic motivations.

The psychological dimensions that drive the motivations of firefighters can provides insight. This can be used to encourage, engage, and maintain the passion of firefighters. Traditionally, from an organizational perspective, this has meant forming cohesive groups of fire officers who provide dedicated leadership to keep the organization engaged.

Recognition and awards are also critical, as is the availability of adequate support mechanisms to help firefighters cope with their jobs' adverse events and hardships. Furthermore, according to Heffernan and collaborators, firefighters are more likely to feel valued and to maintain engagement when they have opportunities to contribute to policy development and decision-making, as well as when they receive formal feedback.[2]

First responders in general face several inherent challenges in motivating themselves—from the public, medical professionals, and patients. These include group dynamics and conflict (e.g., bullying, exclusion, and factionalism), conflicting opinions regarding the scope of practice, lack of support and friction with a health care interface, time pressures, and lack of recognition. Furthermore, lack of advancement or promotion opportunities in leadership roles, balancing personal and work life, and unmet training preferences add to their professional difficulties. Importantly, the stress and emotional turmoil of the firefighter role can significantly impact an individual's life and mental stability at times. Firefighters are frequently required to be on call for extended periods, spend time away from their families, respond to late-night emergencies, be thrust into high-pressure situations, and witness traumatic events.

The Need for Mentorship in the Fire Service

Thus, with so many challenges and the increased burden of being part of the force, firefighters rely on mentors and mentorship programs to prepare for their future. A *mentor* is a person who provides advice and training to a candidate to ensure that the person being trained, known as the *mentee*, receives valuable industry experience and guidance. Receiving this advice and training is the process known as mentorship, typically performed as part of a *mentorship* program.

There is a need for mentors within the fire service, to drastically alter its future and transform its leadership. We must pay close attention to develop mentor/mentee relationships within the fire service. Through mentorship, future leaders are shaped into educated and qualified candidates ready to take the fire service into its future.

Mentorship Programs

According to Bhattacharya, mentees are not the only ones benefiting from mentorship programs. Mentors also benefit through development of their management and leadership skills.[3] Successful mentors can build trust-based relationships with their mentees, leading to a symbiotic working relationship—and often a future friendship. These relationships can also be extremely beneficial, especially if the mentee works at the same station or division. Even if the mentee does not work with the mentor following the program, their paths may cross again.

There are several desirable traits for mentors. According to Andersen, mentors should be able to reflect on past experiences and share advice based on them; these experiences could involve various firefighters and other stations, departments, or states. Both parties—mentors and mentees—also need to be able to have discretion, since mentors may need to reveal personal information above themselves to their mentees to encourage and motivate them. For example, an incident during a fire could have left the mentor with severe physical and mental scars, and the mentee deserves to be informed about the dangers related to the career. In addition, mentors must give honest feedback, so that their mentees can learn from their experiences during the mentorship relationship. Mentors should also have a positive attitude, which will go a long way toward encouraging mentees not to give up when setbacks arise.[4]

A mentee who enters a firefighter mentorship program can benefit from an experienced leader. Hiraki explains that a veteran firefighter can be an excellent mentor to a new firefighter. The veteran firefighter can provide knowledge, expertise, and on-the-job training. Both participants in the mentorship relationship can provide each other with valuable input and interactions that expand their collective knowledge through discussion and completing different tasks together in the workplace.[5]

Hayes asserts that a mentor's job in a firehouse can go far beyond the realm of work.[6] A mentor can build a relationship with an active firefighter that enables them to feel safe and secure to reveal horrifying events experienced in the line of duty. This spirit of connection can also spread throughout the firehouse and create a culture of inclusivity among firefighters by eliminating isolation. Most important, communication lines must remain open so that mentees feel comfortable expressing their emotions to their mentors, confident that their opinions and feelings will be welcomed and heard.

How to Establish Mentorship Programs

Padhi states that a mentoring program should incorporate the following steps. First, the program needs to set goals that are achievable. Depending on the organization, these may differ because fire service careers are complex and vary (e.g., by location and specialization). For example, a firefighter mentor may want to ensure the mentee is efficiently trained in self-rescue. In contrast, a doctor mentor may want to focus on teaching the mentee basic CPR, which could save lives.[7]

Second, the program should offer a variety of topics for discussion to maintain attraction and focus. For example, in the fire service, cases or discussions can include fire operations, different ways to interact with station members, and how to engage with the public. Third, various formats can be utilized in the form of group mentoring. This allows for more than one mentee to benefit from the program. The goal of a successful mentor and mentee relationship isn't to make them more like you but rather to enhance themselves to be a stronger version of themselves. After all, according to Shmidt, mentorship is meant to be "mutually beneficial between seasoned professionals"; therefore, the program should enhance both individuals and allow them to grow in unique capacities.[8]

The mentee is not the only individual who requires support. Difficult situations can arise that mentors may not know how to deal with, and this can lead them to struggle to engage with their mentees. A support system should also be in place in case any challenges arise that are difficult for mentors to deal with. Shmidt states that there should be adequate support for the mentor and the mentee, especially in the form of training. In particular, administrators should help each party to outline their specific objectives, to allow mentees and mentors to understand their roles and interact with others in their journey. Mentorship does not have to be a one-on-one experience—with a support system, the two parties will benefit even more.[9]

Communication is vital to the success of the mentorship program—including in the longer term. Through open communication, mentor and mentee form a strong bond. Furthermore, it is recommended for mentor and mentee to keep in touch. By maintaining contact in the future, both parties can discuss and help other through different experiences on a personal and professional level.

Characteristics of Mentorship Programs

Certain variables need to be considered when creating and implementing the program. One critical element of any mentoring program is for those in senior positions who are charge of creating the program. In particular, they must ensure that the mentorship program meets specific objectives.

Without clear objectives and careful planning, a mentorship program can begin with the best intentions but may not have the same impact as it would otherwise. To ensure that the program meets the organization's needs and goals, senior staff can identify these goals, find strategies or methodologies for mentorship that meet the plans, and then develop a mentorship program that connects these two aspects. Then, it requires constant improvement—even evaluation—to ensure that the program continues to meet the needs of both the mentee and the mentor while aligning with the prescribed goals.

A chain of command should also have designated responsibilities assigned to individuals. Without a clear understanding of who does what, it can be challenging to execute a robust and beneficial program that greatly impacts all involved. Senior leaders must understand the mentorship program thoroughly and then communicate their knowledge and directives clearly down the line.

The mentor and mentee must benefit from the program. Otherwise, the program will be deemed a failure and may not continue. Senior recruitment personnel must carefully screen and select candidates to ensure the program's continuity. Furthermore, they should also keep apprised of the successes and the challenges facing the current participants and the program and make adjustments as needed for better overall success.

Not all employees can be mentors. Candidates who pose a risk in terms of poor past performance or behavior should not be selected. In addition, there may not be enough positions available, and personalities must fit the profile. Mentors should be ethical and honorable people who will integrate well into the program as designed.

Time also needs to be devoted to properly training mentors, to ensure that they can handle the program. Mentors are not born knowledgeable about what it takes to be a mentor; rather, they should be trained and shown examples of influential mentorship practices. How to interact with mentees appropriately should be covered in a training course or workshop.

How to Evaluate the Success of Mentorship Programs

Padhi has identified several ways to evaluate and determine the success of a mentorship program. First, the public safety mentorship initiative should be run efficiently according to a set plan. This plan must allow for deviations, while catching and correcting adverse events to ensure that the plan's objectives are met. Second, the mentees involved in the program need to provide appropriate feedback to make relevant changes. Progress must be monitored within the public safety program in a *personal development plan*. Success can also be further measured in terms of mentor retention rates and the program's sustainability as determined from stakeholder feedback.[10]

Mentorship programs can provide extensive knowledge and valuable experiences for mentees. Mentorship programs that enable future generations of firefighters and fire officers to excel in their line of work are vital to a healthy fire service culture.

Qualified and respected personnel must be appointed to manage mentorship programs effectively. Of Fortune 500 organizations, 71% have implemented formal mentoring programs, and 9 of 10 employees who have been appointed a mentor claim that they are satisfied with their jobs. Furthermore, 67% of organizations with a mentoring program reported increased productivity levels. The future of the fire service depends on well-developed, -trained, and -mentored employees carrying the profession forward.

Managing a Talent Pool

Fire service organizations that have embraced diversity and inclusivity as a demonstrated value can attract and retain a more diverse talent pool. When people feel that they are appreciated and can be themselves—within the constraints of organizational regulations—they will be inclined to be engaged and produce their best work. A diverse and inclusive workforce that is valued for their contributions provides a synergistic effect on the workplace and community.

Chief Niles Ford (PhD) of the Baltimore City Fire Department says, "The reason we exist is to serve our communities." With this in mind, Ford has worked with his community to create selection panels for initial fire and EMS hiring processes to demonstrate his commitment to a diverse and inclusive Baltimore City Fire Department. Community meetings are convened to evaluate eligible candidates to fill the organizational ranks. "The community participates in picking those individuals that best represent the community," he explains. "Once they have made their selections, my training staff will train those recruits to do the job."

This initial step leverages community responsibility with pride of ownership in the city's fire department. Chief Ford believes these actions have demonstrated value to the fire service and community and that this process has been transactional and transformational.

The Gateway to Organizational Talent

One of the most significant changes we've seen in the fire service in recent years is the shift toward prioritizing the complete person during the hiring and promotional processes. While experience, aptitude tests, and other metrics remain valuable, placing too much value on them can create a narrowly focused path that favors some members and excludes others, including certain generations and backgrounds. Technical skills, competence, and proficiency can be taught in a training academy or fire station. Hiring and promoting individuals who reflect the organization's culture and the community can make the difference between an average and exemplary hire. Remember, the initial hire or offer of employment is the gateway to organizational talent.

Navigating this shift involves tackling major challenges, such as entrenched behaviors, embracing new generations, and implementing best practices to supervise members with varied experiences and backgrounds. It's time to reframe traditional supervisory methods if fire service leaders want to create a *bench strength*, or readiness to lead, among their members.

Chapter 10 Review Questions

1. How can mentorship programs contribute to the future of the fire service?
   a. By limiting diversity and inclusivity
   b. By excluding future generations of firefighters
   c. By hindering engagement and productivity
   d. By providing knowledge and experience to mentees

2. Which of the following best describes the joy of leadership when it comes to motivating employees?
   a. A one-size-fits-all approach to management
   b. Evolving and adapting to meet the team's needs and desires
   c. Isolating oneself from the team
   d. Discarding personal aspirations of employees
   e. Forcing employees to conform to organizational objectives

3. How can the impact of a mentorship program be optimized through support?
   a. By ignoring challenges
   b. By offering adequate support to both the mentor and mentee
   c. By limiting communication between mentor and mentee
   d. By avoiding interactions outside of formal settings

4. Why is it crucial for mentors to be ethical and honorable individuals in a mentorship program?
    a. To create challenges for the mentee
    b. To limit the impact of the program
    c. To ensure integration into the designed program
    d. To discourage interaction between mentor and mentee
    e. To hinder the success of the mentorship program

5. Why is it essential for senior staff in the fire service industry to design mentorship programs that meet specific objectives?
    a. To create challenges for the mentor and mentee
    b. To limit the impact of the program
    c. To ensure the mentor and mentee do not benefit
    d. To align the mentorship program with the organization's goals
    e. To discourage mentor and mentee interaction

6. Mentorship in the fire service is not considered essential for preparing firefighters for their future career.
    a. True
    b. False

7. An experienced leader in the fire service can provide knowledge, expertise, and on-the-job training to a new firefighter.
    a. True
    b. False

8. The stress and emotional turmoil of the firefighter role can significantly impact an individual's life and mental stability.
    a. True
    b. False

9. Building a relationship with an active firefighter in the firehouse can create a culture of exclusion among firefighters.
    a. True
    b. False

10. A mentoring program should set achievable goals and objectives as the first step.
    a. True
    b. False

11. What role does communication play in the success of a mentorship program?

12. How can effective leadership evolve and adapt to meet the diverse needs and desires of a team?

13. Why is leadership not solely about leading an entire team but also about developing individuals into leaders themselves?

14. Why is the carrot-and-stick approach not successful in motivating fire service personnel?

15. How can mentorship programs impact talent management in fire service organizations?

11 DEVELOPING COMMUNITY AND ORGANIZATIONAL EQUITY

A good leader is one with a *vision*—a goal encompassing outcomes that can be made to resonate with employees across the organization. Imagine what it would be like for subordinates to have to comply to the demands of the person in charge who has no clear picture of what is happening: The subordinates will be as aimless as the leader; without a leader who understands their vision or lacks one entirely, the subordinates may question or deviate from directives or even go rogue, substituting their own instincts where goals and direction are lacking. A good leader should make possible the achievement of predictable outcomes within the workplace while considering the chance for either a positive or a negative outcome.

As a leader, never confine your thinking and perspective. It is best to always think big and plan bigger! To be a visionary leader, you must be able to see beyond what is on the horizon, to determine the possibilities in any situation. By doing so, you can meet unpredicted challenges head on, rather than living in fear of possible losses. An effective leader must be well prepared for changes to come their way, and they ought to be able to perform their best regardless of any circumstances. A good leader should anticipate possible trouble ahead of their immediate operations and tackle it by managing the potential for loss. By constantly scanning the horizon, exceptional leaders stay one step ahead.

A leader should and must make decisions based on exposure to diverse sexual, religious, and cultural groups. Leaders must think beyond themselves to the population they serve, in both their staff and their community. Being open to different ideas gives leaders room for creativity, to solve problems by thinking outside the box. A good leader must strive to be ahead of their time, with progressive thinking and actions that are linked with the needs of diverse communities.

What's the Difference between Diverse and Inclusive?

The terms *diversity* and *inclusivity* are often conflated. However, contrary to popular belief, these two terms have distinct meanings.

Diversity is often spoken of in terms of metrics: percentiles, raw scores, pie charts, Venn diagrams, and other empirical representations. In a sense, the raw dataset creates the composite sketch of our organizations. It describes the demographic makeup of our organizations and what populations comprise our professional staff and leaders.

As explained in *Managing Workplace Diversity and Inclusion*, when the term "diversity" became prominent in the 1980s, it referred to changes in the demographic characteristics of the labor force and work organizations, particularly race, ethnicity, and sex.[1] In other words, the differences existed between our staff members regardless of the various ways that we could categorize the individuals within our team.

While diversity exemplifies *how* we run our departments, inclusivity exemplifies *why*. In other words, diversity is objective, whereas inclusivity is a subjective reflection of our fire service culture. Inclusion is the spirit of the organization, whereas diversity is more concerned with data. Inclusion is the ecosystem within our organizations that makes members feel equally productive and participate in meaningful work.

Diversity is also the procedural methodology for bringing members into the organization. In addition, as Hays-Thomas writes in *Managing Workplace Diversity and Inclusion*, without a sense of inclusion, employees "may feel like the outsider within; they may experience special stresses, resign prematurely, remain stagnated at entry levels in the organization, or even be terminated." We can analyze diversity all day long, but if we do not make an inclusive environment, our team will struggle and we will fail to provide the support that our team needs to thrive. Consequently, the first step to overcoming diversity challenges is simply to recognize what exists and what differences influence our teams.

Acknowledging the Differences

Differences are natural among diverse groups. For example, simply acknowledging the differences between generations and the challenges they pose is an essential first step in managing a multigenerational workforce. Communication style, negative stereotypes, and cultural expectations are the most prominent challenges. However, they also provide the easiest ways to understand the dilemmas that teams may face.

Communication

Millennials, for example, prefer to communicate with very informal rhetoric through digital technologies and social media. In contrast, Generation X, baby boomers, and traditionalists prefer face-to-face interactions, phone calls, and emails. Understanding these communication preferences can explain workplace dynamics and create the opportunity to find common ground.

Negative Stereotypes

It is commonly assumed that the younger a member is, the less experience they have and, thus, the less deserving they are of their position. This preconception can undermine younger officers' confidence in their ability to do their jobs and cause them to lose faith in their superiors in public safety. By recognizing these harmful stereotypes, we can take the first steps to eliminate these challenges by creating more harmonious and productive staff interactions.

Cultural Expectations

It is also essential to highlight differing values and cultural expectations. For example, between generations, the cultural attributes and expectations tend to contrast much more than people may realize:

- *Traditionalists* are typically straightforward, loyal, discreet, and dependable. They are motivated by recognition, respect, and stability in their occupation.
- *Baby boomers* value loyalty, teamwork, and a sense of duty. They display team-oriented, workaholic, optimistic, and competitive generational traits.
- *Generation X* can be described as informal and skeptical. They deeply value independence, flexibility, diversity, and work-life balance.
- *Millennials* (or *Generation Y*) are civic minded, competitive, open minded, and achievement oriented. They value responsibility, experience, and leadership quality in their workplace.
- *Generation Z* is the most diverse of all. Additionally, they aren't just tech savvy: they are digital natives who seek opportunities to work with technology.

Leaders committed to fostering cohesiveness and efficiency are required in a multigenerational workforce. Leaders must adapt to changing workplace values, attitudes, and demeanors by being flexible and proactive. Leaders in the fire service cannot ignore the changing demographics of the workforce; in fact, refusing to adapt can result in a hostile work environment. Instead, fire service leaders must offer strategies, tactics, and skills that inspire people of all ages. When leaders find ways for their staff to collaborate across generations, the entire fire department benefits.

Benefits of a Diverse Work Environment

Our minds can affect our efficiency and productivity. By adapting to the changes in our surroundings, our minds grow and advance our abilities. If a person's surroundings are filled with cultural, sexual, and religious diversity, this can help them be well rounded.

Exposure to different types of experiences can be gained from traveling. Traveling introduces our minds to new emotions from new surroundings. Exposure to new techniques broadens a person's horizon—enhancing their vision and thought process. This helps them appreciate a larger world view, resulting in a deeper understanding of those around them.

Mindset represents how our brain adapts, for better or worse. Our minds can be tricked into admitting the presumable facts and ordering our bodies to respond—essentially swallowing a placebo. Consequently, the brain sends signals to different body parts, convincing our consciousness that the pill we just took will relieve the pain we are experiencing. At the same time, all it does is satisfy our demand for medicine, which makes it effective. The medicine may not have treated the pain, but our brain successfully convinces us that the pain has been cured.

A healthy mindset in any organization you work with or for is critical to your mental and physical health. This is because we do not take mental health issues as seriously as we should. Not all minds need medical attention, but we must pay attention to how we feel and have healthy ways to cope. We are focusing on our physical well-being every day—which holds immense importance as well.

In the fire service, we are constantly in denial that our mental health is the key to shaping our physical fitness. This cannot be stressed enough: mental health is undoubtedly a pivotal aspect influencing our overall health and thus should not be ignored. You might know many firefighters who go to the gym daily for physical fitness, whereas you might not have come across a single firefighter who meditates or does some mental exercises to improve the state of their mind.

Your mental health is not only important but crucial to your long-term physical health. Studies have shown that those who experience more stress tend to have a higher likelihood of cardiovascular disease.[2] Similarly, mental stresses can have physical consequences such as pain and inflammation. In particular, those with depression or anxiety may experience more chronic pain in the course of chronic illnesses.[3]

To overcome challenges to mental health, there are many ways that you can recover and create self-care practices. You can meditate, perform mental exercises, or spend time in nature. Some people perform breathing exercises, which can undoubtedly benefit your mental health. Health is wealth—that is, both psychological *and* physical health.

The benefits of diverse and inclusive work environments are well documented, including improved problem-solving abilities, creativity and innovation, organizational flexibility, and attracting human talent.[4] More than added value, these are essential to achieving full organizational potential.

Diverse life experiences increase the ability to provide adaptive solutions to technical problems. The fire service should strive for more than diversity based on race and ethnicity, age, and sex, as well as diversity of thought. This last facilitates creative and innovative solutions to move beyond stagnant paradigms and recognize impediments to progress and success to be challenged, changed, and moved beyond.

From Proclaiming Diversity to Promoting Diversity

Internal and external pressures are prompting fire service leadership to expand their rosters and form a more diverse and inclusive organization. Many communities find that declarations of commitment to being diverse and inclusive ring hollow. But this isn't so much a problem as an opportunity.

Only through deliberate and purposeful actions, not empty platitudes, can the goal of creating a culture and workforce that values diversity and inclusion be achieved. If you seek to create a culture and a workforce that is diverse and inclusive, you must be deliberate.

Niles Ford (PhD), the Baltimore City Fire Department's fire chief (retired) has been a champion of diversity and inclusion in the fire service. I spoke with Chief Ford about the importance of diversity and what we as a profession can do to promote it. In Ford's experience, creating diverse and inclusive fire departments has been a lifelong challenge. Nonetheless, rising to this challenge is something that he sees as paramount to the success.

"How can we ever attract people to the profession if they don't see themselves in the role?" Chief Ford says, citing one of the most significant issues. He continues, "I take deliberate action. This may be unpleasant at times, but it is necessary to achieve the desired result. Without deliberate effort, diversity will not emerge."

The fire service's credibility is based on its values and beliefs, including teamwork, leadership, and safety. We must act on those values and beliefs to create diverse and inclusive workforces. If we do not, then the fire service will fail to be as proficient as possible. However, one commonality is critical to the overall success of prioritizing diversity and inclusivity: commitment.

Three Challenges to Commitment

Bridging the gap from the current reality to the desired state requires a commitment to diversity and inclusivity. However, there have always been challenges in how fire service members can ensure commitment from their leaders and within all department employees.

The challenge has been simply stated, but creating and sustaining an effective response can be difficult and complex. Often, the most challenging aspect of this issue is the consistency of the ongoing commitment required. Past efforts have often fallen short for three main reasons:

1. Failure to recognize that the central problem is the presence of a diversity-toxic organizational culture
2. A consequent failure to take a systems approach to meet the challenge of diversity
3. Thinking that the learning curve is flat rather than steep

We must dispel illusions about these factors in our profession. Inconsistent or shifting messaging appears to be at the forefront of these failings. Without proper messaging, consistency of commitment will remain a challenge. When this failing comes from the top, leaders become ineffective.

Changes occur only when leaders believe in and make the necessary changes to reflect underlying values and beliefs. Leaders instill faith and commitment within their organizations, encouraging employees to be dedicated to the cause. However, as we have discussed throughout this book, others can follow only if the leader sets the example. Therefore, the leader must believe firmly to influence others in the department to back up the changes necessary to improve the organization.

Attracting a Strong Talent Pool

The workplace and community benefit from a diverse and inclusive workforce valued for their contributions. People are more likely to be engaged and produce their best work when they feel appreciated and can be themselves—within the constraints of organizational regulations.

Leadership Support Needed

Now that we have discussed diversity and inclusivity and the benefits to individuals, the organization, and communities, let's discuss how to promote and create a more diverse and inclusive fire service. Creating a diverse and inclusive fire department shouldn't stop at adherence to federal law or be solely the responsibility of top leadership. This should be the foundation and from which we work up in increments—in other words, creating a diverse, inclusive organization is a marathon, not a sprint. Once the foundation is stable, continue to adjust to make the most inclusive, diverse workforce you can for your department.

The fire service embraces values-based leadership. Therefore, having our leaders and managers on board with the organizational values and beliefs is—rightly—crucial for any program or initiative. Senior leadership must demonstrate that inclusivity is an organizational value.

Recruitment, Selection, and Hiring

If the effort starts with leadership, the results will appear in the recruiting process. Organizational leadership must collaborate closely with human resources during recruitment and onboarding. It's essential to figure out which underrepresented groups need to be involved. To that end, the fire department should seek out community organizations and schools to recruit qualified candidates. One of the most important steps is to write job advertisements and position descriptions that use inclusive language and images. Work with human resources, marketing experts, and community members to curate the perfect marketing messages to recruit new staff to your department who represent underrepresented groups.

It's critical to train our well-known leaders to be *opportunity managers,* as this is a key role in fostering relationships and ensuring that potential programs or initiatives are aligned with the organization's goals.

You're missing the mark if the internet and social media are your only recruitment tools. Unless you have the inside track on social media algorithms, your fire service organization is probably not visible to half of your constituent community, as Binford explains in "Getting Social Media Recruiting on the Right Track for Diversity."[5] Having a presence at job fairs, trade schools, universities, high school career days, parades, and city- or municipal-sponsored events is a great way to show that your department is diverse and inclusive. The fire service is a unique industry that requires traditional, online, and hands-on approaches to reach stakeholders and potential employees. You are more likely to find staff out in the community than among those searching online job boards.

Selection and hiring practices should come right after the recruitment effort. The selection panel for recruits (or transfers) should also be as diverse as possible. This is a proactive measure to avoid bias, which although natural and expected, can lead to blind spots; your selection panel should be representative of your community since your goal is to recruit a diverse and inclusive workforce. Diverse selection panels mirror how an individual's diversity can benefit the organization and community. More important, they allow your department to remain focused on bringing in the best candidates who may also come from underrepresented populations within your community.

Value Inclusivity and Feedback throughout the Organization

Corporate training is essential for demonstrating and instilling values in the membership. This is also true for building a diverse and inclusive workforce. Demonstrate your commitment to a safe workplace and an environment that values the diversity of race, ethnicity, religion, age, and other federally protected classes, not because it's the law but because it's in line with your department's values. In the end, a culture of diversity and inclusion will gradually grow where it is supported. Opportunities for peer-to-peer, bottom-up, and top-down organizational engagement and community involvement must be taken advantage of.

Encourage people to share their thoughts and feedback. Consider what works and what does not. This starts with asking for input and then having an open, respectful conversation. Incorporating feedback also includes dealing with complaints of harassment or discrimination as soon as possible. Taking prompt, fair, and immediate action in response to claims demonstrates trustworthiness in your stated values. Early intervention can often mean the difference between a teachable moment, a widespread violation or adverse action for the organization's culture, and a tarnished public image.

If you fail to value inclusivity and solicit feedback, you may miss out on critical opportunities. You should not only provide your staff with a chance to contribute but you ensure that your team are all on the same page. With each perspective you gather, your commitment to diversity is demonstrated and will inevitably lead to positive results.

Consistency, Not Intensity, Will Move Us Forward

The statistics show that the odds are stacked against us. As a profession, we are unlike any other in that we are not diverse. The consistency with which we address diversity and inclusivity, not the intensity with which we do so, will make the difference.

When you approach problems and opportunities with purpose, you'll get the results you want in the long run. The biggest challenges are often the ones that go unnoticed—the elephant in the room. Creating a more diverse and inclusive fire service is a chance and a responsibility but only if we look at it that way. My challenge to our profession is to make small, deliberate steps toward becoming more representative of the communities we all proudly serve.

Chapter 11 Review Questions

1. Which of the following is a challenge regarding communication preferences in a multigenerational workforce?
    a. Everyone prefers emails.
    b. There is lack of digital technology use.
    c. Millennials favor face-to-face interaction.
    d. Understanding differing preferences is necessary to find common ground.
    e. Older generations prefer social media communication.

2. Which of the following should be the foundation for every fire service leader and member to create a diverse and inclusive organizational culture?
    a. Adherence to federal law
    b. Compliance with regulations
    c. Shift in organizational goals
    d. Incremental progress
    e. Consistency in messaging

3. How can a healthy mindset in an organization contribute to an individual's mental and physical health?
    a. By neglecting mental health issues
    b. By focusing solely on physical well-being
    c. By emphasizing the importance of mental health alongside physical health
    d. By disregarding the significance of mental exercises
    e. By prioritizing medicinal solutions

4. According to *Managing Workplace Diversity and Inclusion*, what is the importance of inclusivity?
    a. It creates diversity metrics.
    b. It promotes micromanagement.
    c. It increases stress levels.
    d. It fosters a supportive environment for team members.
    e. It encourages premature resignations.

5. Which of the following best describes the process of creating a diverse, inclusive organization in the fire service?
   a. A one-time effort
   b. A marathon
   c. A short sprint
   d. A straightforward task
   e. A responsibility of top leadership

6. Diversity is concerned with data, whereas inclusivity is about the ecosystem within organizations.
   a. True
   b. False

7. Diversity refers to variation in demographic characteristics such as race, ethnicity, and sex in the workplace.
   a. True
   b. False

8. Millennials prefer face-to-face interaction over digital technologies and social media.
   a. True
   b. False

9. Creating a diverse and inclusive fire organization should primarily rely on adhering to federal law.
   a. True
   b. False

10. A good leader should always have a clear vision and plan for the future.
    a. True
    b. False

11. Why is it important to address negative stereotypes in a multigenerational workforce?

12. How do stress levels impact an individual's physical health and well-being?

13. How can a fire department make its recruitment process more inclusive of underrepresented groups?

14. Why is consistency more important than intensity in addressing diversity and inclusivity in the fire service?

15. How can a leader effectively manage differing cultural expectations among different generations?

12 LISTENING AND CRUCIAL CONVESATIONS

That fire you see,
You have it within if only you knew!
This rage you have, it shall take you places.
If only you knew!

—Cormac McCarthy, *The Road*

Communication is the art of giving, receiving, and sharing information. It is an art because an individual's creative skills determine their capability to express information and ideas.

You could have the best idea in your head, but how you explain it to others will determine whether it has a ripple effect on the masses—or make it the greatest of all time! Some are eloquent in their speech; others may have speech impediments, hampering them from articulating a sound argument. While there are several reasons why speech may be impaired, the great news is that these can be overcome with perseverance and patience, allowing you to polish your communication skills over time.

Whether in your personal or professional life, polished communication skills can help you to express ideas clearly and achieve milestones more efficiently. In a professional setting, you must manage deadlines, work on projects, and establish your presence as a valued member of the team—all while maintaining personal health and working on your family and social life.

With hundreds to thousands of people working in an organization, your social circle can expand exponentially. You may meet numerous people whose ideas resonate with yours. However, this depends on how well established the communication channels are within an organization.

Transparency and an open-door policy are ideal strategies. Transparency lends credibility to information communicated, and an open-door policy contributes to peace and harmony in the workplace. On a personal level, the more transparent you are during a conversation, the more people are likely to trust you.

Communication strengthens a relationship more than anything. You get to share your ideas with your teammates and partners, make them understand your struggles, and work on challenges. This maintains relationships in the long run.

Communication and Leadership

As a leader, you share your vision with others, tell them the areas you think could need further work—and mentor your team members efficiently depending on how well you communicate. This helps them grasp the main points of their industry as they work toward achieving individual, team, and eventually organization goals. This is why slashing communication barriers is one of the most prominent challenges leaders strive to address today.

In an emergency response industry such as the fire service, communication is key to successful risk management. You could be on the fireground with a team of experienced professionals, but if there is no harmony among the team members, the entire building could turn to ashes. The power of the team effort is what differentiates a successful, efficient response from one that is lackluster, chaotic, or totally ineffective. The fire service is predicated on a team environment—and by never failing to support each other, we will never fail to help the community as promised.

The fireground is already an intense situation that demands quick decision-making to manage the risks involved successfully. Several lives could be at stake, and not addressing the threats in a timely and effectively manner could result in severe repercussions. Every minute and every second counts. Similarly, every decision and every message we share counts. That's why every fire service leader must prioritize communication.

In addition, communication obstructions on the fireground are unlike those in any other professional setting. While there are general themes—misheard or misunderstood comments, for example—various unique situations make it imperative to get communication right the first time. Firefighters use radios and other technologically advanced equipment to communicate with one another at the fire scene. These tools include features a firefighter needs to avoid as many problems and inconveniences as possible.

Radio ensures the passing of critical information to the right people. In case of emergency, when resources are absent, firefighters radio to inform others of all that is needed, creating synergy among the entire team. Thus, each team stays on the same page, with no one freelancing. With the use of the radio and other tools, we provide everyone tasked with solving the issue and supporting one another. However, some challenges can still arise in the fire service regarding communication.

Operating in a Silo

The Aftermath

Communication and cross-collaboration are typically a challenge in a *siloed* organization. Such organizations have an "us versus them" mindset, within which individuals tend to feel that they are alone and everyone else is against them.

No matter the number of teams or departments, everyone in a siloed organization works in isolation—in their own bubbles. There is no collaboration among team members, and everyone perceives the situation as different, per their own understanding of it. This can create a constant environment of urgency and the perception of threats when there may not be any. By making every man and woman feel like they are in a bubble (or stranded on an island), siloed organizations present a fundamental threat to the unity and teamwork that the fire service relies on.

Siloing causes tunnel vision. Isolation leads the siloed group to operate in survival mode, neglecting outside stimuli and tunneling into their functional areas. This survival mentality has drastic consequences that are difficult to overcome.

In a siloed organization, no one—not one single team member—bothers to work in harmony, making personal growth, not the team's, everyone's focal point. This could be why the insidious effects of silo construction are often not noticed in their incremental parts.

One major problem in a siloed fire service organization is that tasks are duplicated. This is the hallmark of miscommunication or lack of communication. When people cannot decide who receives information directives or assign the job to multiple people, confusion about who should do what occurs. Frequently, other firefighters will attempt to pick up the slack—only to discover that someone else has done the work and they have now completed it again, unnecessarily wasting time and energy, among other resources, to do so.

Additionally, a lack of awareness, energy, or desire to remove silos exacerbates the problem. Moreover, once the groundwork of siloed organizations is laid, the walls go up with little effort—in other words, the problem perpetuates itself. Next, let's look at how to disassemble silos in an organization.

Disassembling the Silo

Disassembling silos is key to stop any further damage and quit perpetuating the challenges that already exist within the fire department. Working in a siloed fire department can have massive repercussions. Yet, since the COVID-19 pandemic, with remote work at an all-time high, more silos have emerged—and many existing silos have deepened. Work disconnects were compounded during this time. With everyone stuck in a bubble, it became increasing difficult to show them that they were not alone in their efforts or their challenges.

At the administrative level, since the pandemic, many fire service officers allowed their employees to work their choice of timings, to retain a semblance of a formal schedule. With regular performance appraisals, they kept quality intact, even when everyone worked from home.

Various communication systems have facilitated these adaptations to the workplace since the pandemic. These gave members the chance to select a form of communication that worked best for their situation and schedule. Some preferred to communicate in writing, while others opted for video calls. Fortunately, opening these lines of communication promoted connection during a time that could just as easily have isolated us. This prevented the creation of a silo, though it remains critical to prioritize teamwork no matter the circumstances.

Working in Teams

Another key initiative is to create—and empower—teams. General Stanley McChrystal calls them "teams of teams."[1] This might not apply to smaller fire departments, but creating teams is essential in larger organizations.

A large fire department will not have individual meetings—and connections—but team meetings. Team members must communicate with one another as they work on projects together. This highlights the importance of staying on task and ensuring that targets are met on time.

Dencentralizing Command Means Empowering Teams

Communication chains need to be created for team members to connect—both within their own group and with other groups. This is called *decentralization of command* or *empowering teams*. There are ways to go about this, but it must be done carefully, to prevent abuse of the decentralized power and control.

Decentralized teams have several benefits:

- They encourage a more collaborative environment that can empower each fire service member.
- When you decentralize, you facilitate efficient, localized decision-making.
- This encourages the creation of specialized teams where individuals can apply their expertise.
- This can also increase collaboration and communication while ensuring that all fire service members maintain a certain level of competence.

Ultimately, empowering teams benefits the organization, as well as the employees.

How a Collaborative Environment Empowers Teams

No matter how it is done, it is essential for organizational members to come together in a collaborative environment. To stop the isolation of people within silos, senior leadership teams must explore and be open to new ways to connect and communicate with one another.

This is one reason I started my mentoring program. The program helps a leader connect with a diverse group of individuals, staying updated and informed of challenges that may arise. They can then determine how a firefighter should best deal with them. This is a decentralized method, because it allows specialization. This benefits the individuals and the group through increased efficiency and the sharing of unique expertise between mentee and mentor.

Using online platforms such as Zoom or Microsoft Teams for video meetings and chat conversations could be ideal for people to share and learn about their tasks, supporting one another during crises in a collaborative environment. With technology advancing into newer realms and with the introduction of newer gadgets, tools, and devices, organizations may use it to silo their environment. Nonetheless, in my department, I have seen firsthand the benefit of digital tools and believe that they have excellent potential for future communication within the fire service.

Strong Communication in a Fire Service Organization: How Do We Do It?

In the fire service, we develop training and methods for people to share, learn, and improve their emergency response skills. One of the most generic ways to promote connectedness, rather than siloing, is to encourage and incentivize the team. Many fire service organizations don't follow this approach, and it is shocking how many departments could better serve their communities and personnel if they chose to emphasize teamwork.

Integrating teams within an organization can be done no matter the size of your department. A team atmosphere is as easy as setting goals for the organization and fire department leaders. If people feel they are all working toward something together, they feel more connected and purposeful in their obligations.

Suppose firefighters are unsure where they might be failing. In that case, they should be supported by their leaders. Their concern, after all, encourages a review at the department level to discover the specific loopholes or other failure points affecting the team's productivity.

Can you Hear Me? How Communication Can Be the Weak Link

Poor communication could wreak havoc on the fireground, especially when firefighters try to extinguish the fire. In fact, communication is often the weak link, whether on the fireground or in the firehouse. It is critical to send and receive clear messages to ensure a successful fire operation.

You may also figure out if you are talking *with* people or *to* them. This could also be evident in their response and actions. How they respond to you will help you recognize if you are communicating well or need to take steps to improve your approach.

Firefighters often have to ask several times if the concerned person heard them correctly—or at all. "Can you hear me?" they ask, while fighting for people's lives and putting out the fire. Even though the answers could vary depending on the party listening to you, how you have delivered your message is critical to successfully communicating your point. The communication equipment you use could also make or break the ongoing collaboration in a stressful situation.

Ways to Communicate Well

Communicating is the first step, but you must also communicate well. There are a variety of mistakes that can make communications less viable—from being unclear to not listening. Strong communication means getting the point across to others.

Here are a few tips to strengthen your communication:

- Speak with clarity.
- Explaining everything well, so that your counterpart can understand without question (which may not be possible).
- Match the frequency of the people you interact with, in pace, tone, and energy level (not only will this make them feel more comfortable, but it will also heighten your connection and lead to more profound understanding between both parties or even your entire team).

Active listening is the way to go, whether in the firehouse or on the fireground. Because communication is a two-way street, listening helps us to decipher relevant information efficiently, to understand a situation and make decisions in a timely manner. Active listening means paying undivided attention, avoiding judgment, and demonstrating empathy. You should ask clarifying questions and summarize or paraphrase what you have heard to ensure that you have fully and correctly understood the other person's message. This will ensure that you understand them and that they are more likely to return the favor and be fully present when you share your feedback or thoughts.

On the fireground, we must be attentive, understand the situation, and make timely decisions. Listening to what is happening around us and what others are saying helps us to actively participate in putting out the fire.

A final point regarding active listening is to put the gadgets down. Only this way can you participate in the conversation and truly listen to what others are saying. Not just on the fireground but also during project planning, listening allows us to brainstorm innovative ideas and successful strategies in risk management.

Comprehension, Compassion, and Empathy: The Keys to Strong Communication

In short, three main attributes make up strong communication: comprehension, compassion, and empathy. Without comprehension, you cannot proceed. You also need compassion and empathy for others to ensure you can effectively communicate, to actively listen and be heard, and to comprehend or decode what is being conveyed.

This also has implications on how different generations communicate. To share their messages, the generations must cross barriers to understanding each other's perspective. In practical terms, to engage members across all generations, leaders should learn their members' preferred method of communication (e.g., in-person, phone, email) and encourage such use among the team—and by themselves when communicating with those same individuals.

In general, while older generations expect and enjoy face-to-face communications or picking up the phone to talk, millennials are more likely to send the call to voicemail—and they may not even listen to their messages—and respond via email or text, and Generation Z will default to online meeting and digital messaging platforms.

Through these insights, we can encourage better communication while ensuring that we remain empathetic and compassionate about the situations and issues affecting people. If you master and apply all three of these traits, your communication will be not only sound but also impactful for your organization, especially when it comes time to engage in hiring, onboarding, and training your newest department staff members, as we'll discover in the next chapter.

Chapter 12 Review Questions

1. What is the default mode of communication for Generation Z?
   a. Sending emails
   b. Speaking face-to-face
   c. Using messaging platforms and hosting digital meetings
   d. Avoiding all forms of digital communication

2. Which form of communication do older generations tend to prefer most?
    a. Sending emails
    b. Using messaging platforms
    c. Hosting digital meetings
    d. Speaking face-to-face
    e. Ignoring team members' preferences

3. Why is actively listening important for strong communication?
    a. It helps in ignoring relevant information.
    b. It ensures misunderstanding of the situation.
    c. It aids in decision-making and understanding the situation.
    d. It encourages judgment and interruption.

4. Which of the following statements best describes communication?
    a. It is solely about speaking eloquently.
    b. It involves giving, receiving, and sharing information.
    c. It is important only in personal relationships.
    d. It is unnecessary in a professional setting.
    e. It is a skill that does not require polishing.

5. How can communication skills help leaders to achieve milestones efficiently?
    a. By struggling to communicate ideas
    b. By isolating them from the team
    c. By maintaining a closed-door policy
    d. By establishing a sound argument

6. Putting aside gadgets and actively participating in conversations can aid in effective communication, in both firefighting and project planning.
    a. True
    b. False

7. Encouraging a team atmosphere in a fire service organization is not necessary for improving emergency response skills.
    a. True
    b. False

8. Using online platforms such as Zoom or Teams could be ideal for people to talk, share, and learn about their tasks, supporting one another during crises in a collaborative environment.
   a. True
   b. False

9. Active listening is an essential component of strong communication as it helps in understanding situations and making effective decisions.
   a. True
   b. False

10. Strong communication in any organization primarily relies on three main attributes: comprehension, compassion, and empathy.
    a. True
    b. False

11. What are some key considerations for organizations looking to create a communication strategy that effectively addresses the needs of diverse generations in the workplace?

12. What are some consequences of operating within a siloed organization that lacks cross-collaboration?

13. In what ways can setting goals for the organization and fire department leaders promote a team atmosphere in a fire service organization?

14. Why is it important for a fire service organization to prioritize team cohesion and communication even during challenging times when remote work is necessary?

15. How can understanding different generations' preferred methods of communication improve overall communication within a team?

13 HIRING, INTERVIEWING, AND ONBOARDING

Ideally, employees will understand the vision, mission, and values of an organization because they work toward its goals. An organization could have everything aligned—from leadership to policies and working procedures—but if employees do not know why they do what they do, it is bound to fail. The human resources are the most valuable assets of any organization because they can make or break it.

Tremendous effort goes into finding the right people for the proper role because not everyone is an ideal fit for every particular job. Given that the fire service is already a tough industry, job recruitment is inherently challenging. Recruiters cannot simply hire anyone interested; rather, they need to determine that their efforts to onboard and train the new hire will not waste the organization's time, energy, and budget.

Several factors come into play when deciding which candidate to hire. Employers look for individuals from a large pool of candidates seeking the job and, after a thorough process, choose to offer them the job they have applied for. This process usually involves identifying, screening, short-listing, and hiring potential *resources* to fill vacant spots in the organization. In technical terms, this is called *recruitment*, or *selection*. As a human resource core function, recruitment entails choosing the right person for the right position at the right time.

In any fire service organization, recruitment determines how effective its plans will be during emergencies. Recruitment and retention are the two most critical procedures in any fire service, especially when recruiting volunteers.

Multifaceted Approach to Recruitment

Fire service recruiters have forever been looking for ways to attract and retain members. They have endeavored to develop strategies to recruit skilled staff before their organization becomes understaffed and the existing employees feel burdened.

In many fire departments, one challenge recruiters face is burnout and employee retention. After all, for every single position that needs filling, there are hours of work being covered by the department in the meantime. For example, in the Tucson (AZ) Fire Department (TFD), several vacant positions were available to hire new staff, and the existing firefighters felt the strain. Even though TFD believed that they had a low vacancy rate in terms of numbers, firefighters were required to work overtime, leading to burnout. Therefore, TFD took a multifaceted approach to recruiting: They revamped the fire department's recruitment page on their website and targeted a pool of various skilled professionals.

TFD also began recruiting through different channels, not just by word of mouth. They used social media platforms to bolster their recruitment efforts, with great results.

In addition to taking a multifaceted approach to recruitment, most fire departments have strategized their recruitment processes as described in the following sections.

Increasing Current Employees' Pay

Instead of offering a signing bonus for new applicants, you could redirect those funds into rewards and pay increases for current staff. This will help retain the existing team, and retention in turn will save cost and time that would have been spent training new hires. With bonuses and pay increases, your current staff will feel more committed and likely see reason in remaining with your specific department instead of transferring to another department—or worse, quitting.

While one-time bonuses undoubtedly improve existing staff's motivation, the impact on retention could be magnified if employees moved from receiving living to thriving wages. Even with the importance of the other issues discussed throughout this book, retaining staff and ensuring that you have sound, committed personnel within your department should be at the top of your list! The happier your employees are, the better your team will be—and the more likely you will attract people who want to become part of your exceptional organization.

Recruit All the Time with the Best People

Everyone in the department must take recruitment seriously. Moreover, you need to recruit the right people at the right time. Thus, recruitment should be purposeful at every step of the process—from the communication with candidates and the interview process to training and onboarding. All of this has one goal: getting the perfect candidate into the available role.

It could be argued that optimism for the future of EMS is low. Several years of EMS Trend Report data indicate that only a quarter to a third of EMS professionals are likely to recommend a career in EMS to someone they know.[1] This is alarming for several reasons.

Those who are optimistic about their future in the fire service or EMS and confident about the profession's future are likelier to attract others to join their organization.

Alarmingly, though, such optimists may feel less committed to their jobs, less fulfilled, and in turn, more likely to quit. This is another issue to concern yourself with—as its resolution will be critical for the success of your department. However, to remedy this concern and ensure that you find the right people, turn off the devices and focus on people face-to-face.

Interviewing Applicants Face-to-Face

Since the COVID-19 pandemic, organizations have incorporated newer methods to attract interested candidates. Offering work from home and interviewing candidates online are two of these.

After years of grinding isolation amid restrictions and precautions for gatherings, many individuals yearn for face-to-face communication. They seek platforms where they can speak up and interact without worrying about social distancing. The pandemic also introduced unique ways to conduct interviews, and these have been effective for some industries. However, conducting face-to-face interviews appears to be more effective for EMS and fire service.

This is why most recruiters consider going where they will likely meet potential applicants. A fire service recruiter might set up a booth or table at a local farmer's market, music festival, house of worship, or other places people gather. TFD, for example, set up a recruiting booth at community and sporting events, as well as at military outplacement job fairs.

Recruiters must be tenacious. They might need to travel to a neighboring state or learn more about high school and college sports in their community, to reach and engage with a broad candidate pool. These are several ways to recruit face-to-face, and it is important to do so. Through meaningful connections and impactful introductions, people will gain direct experience and understanding of what your department—and they—can offer.

Leadership experiences and stories of performing under pressure might inspire and educate other recruits who are genuinely interested. For many people, meeting EMS or fire service professionals in public is a meaningful experience that has more impact than any internet research on the job of interest. At the same time, these leaders could even use those connections to share opportunities in public safety with people who could be physically fit, have performed under pressure, or seek coaching to polish their physical and mental performance. There is a dual purpose to becoming involved in the community and engaging in face-to-face recruitment practices and interviews.

Go Beyond Social Media and Web Advertising

You should use social networking, a department website, and web advertising to attract applicants. Instead of simply shouting into the void of cyberspace, focus your attention on the people most likely to be interested in a public safety career. Start by targeting messaging at the family, extended family, and friends of current personnel and retirees. These people already have a foundational understanding of the career and may want to follow in the footsteps of someone they admire and respect.

Beyond your initial contacts, begin small and work out within the community. The first step will often be to set up a booth at a community event and share more about the opportunity with those who come forward. Remember: firefighters are driven by their pursuit to protect and serve their community. What better way to find the people who may be most passionate about defending their community and protecting those they reside nearby than by going beyond the internet and directly into the community?

No matter how you go about your marketing and recruiting, make an extra effort to be precise regarding your target market. After all, not everyone can be a firefighter, and you cannot accept just anyone to your department to integrate into your existing team.

Be Clear and Precise on Who You Want to Apply

Every time EMS1, FireRescue1, or Police1 have asked current public safety professionals' reasons for their services, the most common responses have always been to serve their community and make a difference in the lives of others. Before launching any recruitment campaign or posting a help-wanted advertisement, you must seek answers to the following:

- Will this be seen by people who want to serve and make a difference in our community?
- Does this communicate how we serve and make a difference in the lives of the people in our community?

If you answer no to either question, you must retool the message. Refine your efforts to reach your target market and find the exact type of person you know will excel and want to be part of your team.

Question Everything and Anything about Retention and Recruitment

When it comes to retention and recruitment for EMS agencies and fire departments, changes are always ideal. You must continuously polish, reform, and innovate to develop better recruitment strategies.

Question everything about your retention and recruitment efforts. If specific processes seem to be working, continue using them. If not, then innovation might not hurt.

For example, you could question your staffing models and shift lengths. Or brainstorm ways to adapt the job to the current needs and interests of the available workforce. Ultimately, recruiters should determine if their methods are effective and adapt if not.

Evade Stress and Fatigue

What levers can you pull to reduce stress and fatigue for individuals working in your organization? How could you make the job more appealing to those you want to welcome into your industry? These are questions recruiters must answer to interest candidates in applying for the job you are offering.

As a recruiter, find your reasons for hope and optimism about the future of your organization and the profession and install them in the applicants' minds. Many things come in cycles, and this cycle could be pretty challenging.

Finding and Recognizing Talent for the Fire Service

Creating a diverse and inclusive fire organization shouldn't end with adherence to federal law or be solely the responsibility of top leadership. Remember to work in increments; creating a diverse, inclusive organization is a marathon, not a sprint, so it could take time to find talent to perform in the fire service.

It is critical that once an individual is hired, they are treated with the utmost respect. Further, the organization should provide training and development programs so that the staff meet the necessary proficiencies.

Training helps employees understand the procedures aptly and perform in the manner expected. However, trainers should recognize that everyone has a unique skill set that distinguishes them. While some employees could be strong at analytical skills, others could be exceptionally fit physically; some could be fast at handling technical matters, and others could be strategic thinkers, having solutions to every problem.

Recognizing talent is also critical. Use your employees to their actual capacity and the benefit of the organization. Reward them for efficiently using the existing procedures. That is, the more the individual's proficiency levels are identified, the more efficient the organization's performance will be in reaching its goals.

Command Presence and Resonance

What is Command Presence?

Command presence is whether individuals can be respected for their attitude and management of a situation, mainly based on their high level of efficiency and assertion. As a long-time participant in the fire service, I have consistently found challenging the idea that command presence indicates the true leadership potential of candidates. Nonetheless, in every promotional process or assessment center I have participated in, the grading criteria have always included whether the candidate had command presence. However, I find this to be a supremely subjective idea, especially because it's difficult for many people to define this term. Command presence is one of the things that we see more than we can define or quantify.[2]

Understanding Command Resonance

My call to action is to define another leadership ethos, which I call *command resonance*. The leadership of our profession must embrace a departure from the strict teaching of and desire for strong command presence. If we seek and train for this, we can also become resonant leaders as well.[3]

Many people are unsure what it means to be a resonant leader. Resonance entails understanding people's feelings and emotional state and allowing them to move in a positive direction. Resonating with the tone and tempo of emotions enables people to connect with those around them and uplift them. It may sound easy to inspire people, but it can be tough to achieve resonance with diverse emotions across an organization.[4]

According to Goleman, through resonance leaders can demonstrate responsibility while also affecting a substantial amount of organizational performance.[5] This requires being comfortable with being uncomfortable. However, working beyond your comfort zone is a critical challenge for many people, regardless of their profession since this issue also exists outside of the fire service; nonetheless, fire service officers and chiefs need to understand what if any qualities of command presence they can exhibit.

The real question is, do we need these leaders to demonstrate command presence? The answer is always yes. Given that this is invaluable in emergent environments, this developed skill is needed approximately 5% to 15% of the time; the rest of the time, the leader needs to exhibit command resonance.[6]

Command resonance creates positive feelings and outcomes using emotional intelligence, also called *emotional quotient*. Successful implementation and development of command resonance throughout an organization can yield better emotional stamina, increased collaboration, idea sharing, and even formidable emotional bonds that allow the organization to move through any crisis. These are undoubtedly crucial to most fire departments when they are on the job, whether in the firehouse or on emergency scenes.[7]

Command presence is critical for performing the full scope of what it means to be a fire officer. Retired battalion chief Frank Ricci's book *Command Presence: Increase Your Influence* presents a comprehensive guide on leadership, resilience, and personal accountability, drawing heavily from the author's extensive experience in the fire service. Ricci's book provides practical lessons and insights for leaders across various fields, focusing on how to effectively handle adversity and build a strong command presence.[8]

Overreliance (even accidentally) on command presence, as with any leadership ethos, can be problematic. In particular, it may prevent people from speaking their opinions because of fear of repercussions. Still, team members might not entirely trust a leader who does not exhibit command presence, given that they may be concerned with avoiding risk. Command presence requires leaders to be in sync with the members of their teams and their emotions; moreover, prevails when immediate action is needed, where bonds of command resonance would require too much time and energy to develop. Command resonance, by contrast, takes time and energy to cultivate.[9]

Characteristics of Command Resonance

Command resonance is ultimately the key to creating a high level of engagement and a solid organizational culture. So, now comes the question: How does a leader develop command resonance? It is not easy, as it takes consistent effort and desire, but five characteristics illustrate how and when to recognize and create this command resonance in your staff or yourself.

Self-Awareness

Self-awareness, as the first important trait required for command resonance, helps people to actively participate in their emotional states. It is not essential to label your feelings as good or bad but rather to become aware of how you feel; in other words, it is better to sit in emotional discomfort than to avoid feeling what you feel. This pillar of emotional intelligence allows you to become aware of and manage your emotions better.

One essential tool is a journal. Journaling allows you to check in with yourself and to provide internal feedback. It is best to do this regularly, to progressively grow and understand your emotions or recognize common or constant roadblocks. These should be seen as growth opportunities prompting you to take action(s) to develop your self-awareness and become a better individual and a better leader.

As a firefighter or fire officer, it is essential to understand yourself when under stress. Introspection, or self-reflection, is not about weakness but about becoming more available for personal management. You need to understand who your genuine self is by taking time to reach this understanding honestly and compassionately and then further examining the inner and outer worlds that you exist in. These questions allow you to become self-aware and move into a command-resonant leadership position:

- What mistakes do you make?
- How often do you make them?
- How do you handle them?
- What are the things that you need to work on?
- What makes you the proudest?
- How do you respond to crises?

Self-Control

People with *self-control* can compose themselves, understand their feelings, and manage their impulses. They can work through or dispel distressing feelings. They stay composed and positive even in the most difficult of times.

Self-control is a necessary trait for a command-resonant leader. Often overlooked is the ability through self-control to sleep on a decision or a situation. Whereas many people might act on the compulsion to tell somebody off or write a scathing reply to somebody, a command-resonant leader holds off; writing down your thoughts can be cathartic, but deciding whether they should be aired is ultimately an application of self-control. Having constant self-control and self-awareness in various situations makes a leader disciplined and principled.

Self-control is also about looking at your commitments and promises, allowing you to ensure that you are accountable for your decisions, actions, and outcomes. You need to own up and admit your mistakes, allowing you to gain trust, reliability, and authenticity as a leader and individual. You can become an adaptable leader with self-control, demonstrating how you handle multiple demands and shifting priorities in a rapidly changing environment. Those who work toward this more resonant state of leadership adapt their responses and become more flexible, ultimately developing an essential skill set in the field of fire service.

Social Awareness

Social awareness relates to how everybody interacts and treats those with whom they are in contact. Social awareness and relationship management are critical on the path toward resonant leadership. Simple acts go a long way. Greeting somebody by their name, not interrupting others, or recognizing significant dates critically influences and indicates social awareness.

Other related skills are worth practicing, too. As discussed in chapter 12, use active listening and be fully present when having conversations. Additionally, responding to people who have a question or suggestion can be used to develop buy-in and let them know that their opinions are valued.

Social awareness comprises understanding the social cues you engage with regularly, both verbal and nonverbal. *Verbal cues* are the words people use and what they mean, often both direct and indirect. Nonverbal cues are imparted by body language and can communicate much more than you might realize. Practice interpreting body language to recognize the emotional state of others. You can even become more aware how to use your own body language.

As an example of a nonverbal cue, one of the most off-putting, socially impotent behaviors is checking a watch or phone several times during a conversation; this demonstrates that you do not have the time or are not engaged in the conversation. Another nonverbal cue is how someone either faces you during conversation or evades your gaze, demonstrating that they are not participating or maybe even hiding their thoughts. By contrast, making direct eye contact and facing the person you are speaking with are nonverbal cues indicating that you are actively listening; this is level of the social awareness to look for in new talent and curate within your team.

Social awareness and relationship management practices foster a positive work climate by making employees feel comfortable. If you can master social awareness and find others who also can, then you will see a significant shift in your organizational culture and enjoy a greater degree of efficiency together.

Competence

In short, *competence* is the ability to regularly undertake responsibilities and perform to the organizational standard. It's a combination of skills, experience, and knowledge. Competence in the fire service begins with unifying experiences and dedication to the craft.

For new firefighters, it can take time to become competent. However, your prospective recruits should be somewhat competent and able to handle the job. Technical training, education, safety protocols, problem-solving skills, and teamwork, as well as physical fitness, are all prerequisites. You should prioritize finding individuals who can demonstrate these abilities and are willing commit to continually increasing their competence throughout their career.

Being competent in the job leads to expanded opportunities such as leadership positions. This is where the path to building command resonance will be more crucial. In a perfect world, you should strive for complete competency to continue the team's excellence and promote within your team once the time is right and the position is available.

Authenticity

The most potent and desired characteristic of the command-resonant leader is *authenticity*. You want to hire and recruit command-resonant leaders who are genuine. In other words, you want people who walk the walk, not just talk the talk. If one thing is sure, the fire service can spot a fake. To accentuate authenticity in oneself, you must exhibit the following four characteristics.

The first of these is *acceptance*. The better you know and love your authentic self, the less inclined you will be to judge others. As John Mark Green writes, "The self-righteous scream judgments against others to hide the noise of skeletons dancing in their closets."[10] In yourself and the people you hire, strive to demonstrate acceptance, encouraging each other to love themselves as they are and always keep an open mind about the world around them.

The second is exhibiting a high level of *integrity*. Authentic people live by their values, even when doing so is controversial or unpopular. As discussed in chapter 6, integrity can make the difference between your team understanding your values versus having difficulty seeing the end goal. When you practice integrity, your team follows your example.

The third characteristic is *courage*. It takes great courage to take an honest look inward and discover who you are. It takes even more courage to present your genuine self to the world. In the fire service, leaders must be courageous in the line of duty and in pursuit of critical solutions. Also prioritize courage during recruitment: you want to hire those who can walk into the fire to save others, so they must demonstrate that they are brave and capable.

Finally, authentic leaders who have command resonance exhibit *forgiveness*. They recognize that character flaws, mistakes, and even hurtful behaviors are almost always unintentional. People who practice forgiveness in their own lives are more inclined to allow others the grace to do the same. In yourself and your potential hires, forgiveness is essential to ensure a sound, positive team environment.

Faith and Trust

Command presence will reign supreme in the arena of emergency operations. It is my hope, however, that we seek the opportunity to become more resonant in our leadership capacities. Creating a positive, vibrant environment allows us to have faith in each other and trust the purpose, actions, and intent of others to further the mission, vision, and values of our organizations.[11]

What Happens after Hiring: The Onboarding Process

After conducting your interviews and hiring suitable candidates who fit these traits, it is time for onboarding. Onboarding in the fire service is about ensuring not only that they are well apprised of the rules of your organization but also that they integrate well into your existing culture. As mentioned in chapter 2, a positive, mission-driven organizational culture plays a crucial role in this effort. Still, there are other aspects of the onboarding process to consider.

The long-term retention of your staff members—new and old—is rooted in two training processes: continual training and immediate training. Your onboarding process should not only be comprehensive but also include clear expectations. Outline the culture and let your new staff members know what to do. Not only does this allow them to understand the organizational culture, but it also gives them a chance to integrate easily into the existing, proficient staff you have on board already.

If you follow a sound hiring and interviewing process—keeping in mind all that has been mentioned in this chapter—you should have no issue with onboarding and, eventually, your success. From one leader to another, the tone you set in the hiring process is crucial, so make sure you allow your potential new staff members to be inspired, acclimate, and excel in their new work environment. If you master organizational culture and lead by example, this will likely be easier than you expect.

Chapter 13 Review Questions

1. What is the primary goal of training and development programs for staff members?
   a. To discourage problem-solving skills
   b. To inhibit the recognition of talent
   c. To prevent emotional intelligence
   d. To train employees to perform in a way that is expected of them
   e. To minimize efficiency in organizational performance

2. What is essential for the long-term retention of staff members, both new and old?
   a. Offering high salaries as incentives
   b. Providing sporadic training sessions
   c. Setting unclear expectations
   d. Providing both immediate and continual training
   e. Restricting access to existing staff members

3. Why is it important to outline the organizational culture during the onboarding process?
   a. To overwhelm new staff members with information
   b. To ensure the focus is solely on technical skills
   c. To help new staff members integrate easily with existing staff
   d. To create a competitive environment among team members

4. What is the primary goal of the multifaceted approach to recruitment in fire departments?
   a. Increasing pay for current employees
   b. Retaining existing staff
   c. Hiring new applicants
   d. Focusing on word-of-mouth referrals
   e. Attracting skilled professionals via various channels

5. Which of the following best describes the ideal focus of recruitment efforts?
   a. Targeting random individuals
   b. Recruiting solely through the internet
   c. Attracting people who enjoy social distancing
   d. Finding individuals passionate about defending their community

6. Self-control is not necessary for becoming a command-resonant leader
   a. True
   b. False

7. Employees who are optimistic about their future in the fire service or EMS are more likely to attract others to join the profession.
   a. True
   b. False

8. Self-awareness is not a crucial trait for developing command resonance.
   a. True
   b. False

9. It is important for organizations to align their recruitment strategies with their long-term goals and objectives.
   a. True
   b. False

10. Admitting mistakes and being accountable for decisions helps a leader gain trust and authenticity.
    a. True
    b. False

11. How does practicing forgiveness benefit team dynamics in a leadership role?

12. What key factors do recruiters consider when hiring for a tough job such as firefighting?

13. What are some examples of locations where fire service professionals could set up recruitment booths?

14. What can fire service recruiters do to target messaging at potential recruits in the community?

15. What role does authenticity play in command-resonant leaders?

14 STOICISM, RESILIENCE, AND THE FIRE SERVICE

A blazing fire makes flame and brightness out of everything that is thrown into it.

—Marcus Aurelius

By now, you may have realized that carrying the fire is not as easy as you had imagined. Being a part of the fire service requires something else of you, something not present in many other professions. As firefighters, we are a different breed. We are the ones who run to the danger to save others, rather than run away. In many ways, we are protectors, heroes, and leaders; thus, we are cultivated with different expectations.

There are various ways to describe us in our profession and who we become over time. Two terms in particular are used to describe firefighters and their attributes: stoic and resilient. Without burying ourselves in philosophical ponderings, let's explore each of these individually.

Stoicism and the Fire Service

Fundamentally, *stoicism* is about cultivating a rational approach. This entails recognizing what is within one's control versus what is not. Digging deeper, it is about acceptance.

The classic principles of stoicism set forth by Seneca the Younger, Marcus Aurelius, and Epictetus are straightforward and concise. In short, stoicism is leveraged by four virtues:[1]

- *Courage* is the mental strength to endure, persevere, and overcome danger or adversity. Courage is built on experience and competence. The opposite of courage is fear, dread, and anxiety resulting from uncertainty and inexperience.

- *Wisdom* is the accumulated appraisal of insight, judgment, and knowledge. This is not fixed but something that continues to grow as the individual is exposed to more complex, often critical, life events and challenges.
- *Self-control*, or *moderation*, is the ability to regulate oneself. Specifically, this means controlling one's emotions and desires versus expressing those same emotions and desires in one's attitudes and actions. This is amplified in adverse situations.
- *Thriving in adversity* means that we should not avoid hardship. Instead, we should adopt a more competitive attitude toward our challenges. Anticipate being uncomfortable but prepared. Narrow your attention to the task at hand. Focus your energies on those activities that will produce the desired result. This is how one thrives on adversity.

Excellence withers without an adversary.

—Seneca the Younger

These four virtues are a road map for the fire service. The nature of our profession imparts the need to understand, embrace, and implement stoic virtues. We cannot control where, when, or what the emergency response may be. While we may not know the conditions of the event, such as the rate of fire spread, number of victims, condition or construction type, and inexhaustible list of variables that pose challenges in every emergency, we can, however, apply the four stoic virtues to every incident and everyday task.

Stoicism allows one to maintain an indomitable spirit in the face of irrational or unfortunate circumstances and, in turn, overcome those hardships. The stoic does not overreact. Challenging traditional responses becomes habit with practice. We often cannot change outcomes, but we can change how we respond to them. Stoicism reengineers negative opinions and emotions into a winning mindset that features perspective and readiness. With a stoic mindset, obstacles and challenges become opportunities. Modern stoics practice training one's perception by turning the obstacle upside down. As Aurelius famously said, "The impediment to action advances action. What stands in the way becomes the way."

Stoicism is a growth opportunity for fire service professionals—from recruiting firefighters to fire chiefs. Discerning what is within your control and what is not lies at the heart of everything we do. According to the stoic, nothing is truly good or bad, only your perception. Therefore, we become better opportunity managers if we challenge our mindset to seek opportunity in adversity.

The more we value things outside our control, the less control we have.
—Epictetus

Any experience beyond one's choice should be seen as a chance to demonstrate strength and mental fortitude. The mental framework is to see the circumstance not as misfortune but as opportunity. Practicing stoicism differentiates what we can exert influence over from what is immovable so that we spend our time and efforts on those things we can change.

The following are the three desired effects of practicing stoic principles for the fire service:

- Becoming an opportunity manager
- Increased mental, physical, and emotional fortitude
- Increased confidence

Stoicism develops firefighters' talents and abilities to work well together. Essentially, it provides a way of committing oneself to excellence. As we hire and train the profession's newest members, stoicism enhances their problem-solving abilities and instills confidence in their aptitudes. Stoic firefighters also develop their professional integrity: they do what is right not to impress others but to impress themselves.

For first-line officers, lieutenants, and captains, stoicism enhances their situational awareness by improving their recognition of obstacles with a cue-primed response model based on confidence, not fear. Further, by exercising the principles of stoicism, officers will become better opportunity managers: They will learn not to attempt to control all aspects of the firehouse or fireground chaos but rather to seize opportunities and address challenges from a confident, self-controlled perspective. Officers who realize that they do not have to have all the answers but who maintain a desire to learn and integrate from other examples will exhibit the stoic principle of wisdom.

For chiefs and administrative officers, their years of practicing the art of fighting fire will coalesce with the principles of stoicism. And yet, learning and practice are never complete. The goal of self-mastery and purposeful action guides the executive officers' decision-making processes. The chief officer must learn and demonstrate being dispassionate meetings and other events that elicit strong, visceral responses. Again, not every action requires an emotional reaction (see chapter 13's "Self-Control" section).

Additionally, in their administrative roles, chiefs and officers have the opportunity to be responsive and demonstrate equanimity to political and organizational fluctuations. The stoic recognizes their weakness and works with deliberate action to address them. This growth allows chief officers to discover and develop an appreciation for diversity of perspectives and opinions.

Resilience in the Fire Service

Aside from stoicism, which offers discernment and the opportunity for balance, *resilience* is essential to any firefighter and the ever-evolving fire service culture. This is the process of adapting well in the face of adversity, including trauma, tragedy, threats, or other sources of stress. This can come from family, friends, relationship challenges, serious health problems, financial challenges, or workplace conflict. How do you strengthen your resilience? It's simpler than you think.

First, connect with others and learn about their challenges. Empathize with them and share your feelings with others. Prioritize the people who you care about, but also care for yourself. As discussed in chapter 8, you must take care of your body. Food is fuel. Your mind should also be cared for, so avoid negative behaviors and deal with your stress, rather than creating a mountain of problems that you will still need to deal with later. Employ mindfulness practices and watch your mental health and overall resilience soar.

Remembering your *why* is another vital way to overcome any challenge and build resilience. What brought you into the service? How did you come to be a firefighter? Why do you get up in the morning and go to work? Remember your purpose and why you do what you do. This is what builds your resilience and, ultimately, what builds up the grit that you need, as Duckworth writes in *Grit: The Power of Passion and Perseverance*. This is one of my favorite books and integral to our discussion of resilience. Duckworth believes—as I do, too—that your endurance and your resilience are more important than your talent, allowing you to persevere.[2] You can be the best firefighter in the world, but if you can't persevere, you won't get anywhere. If you think you don't have grit, here's the best news: you don't need to be born with it; you can develop it.

Just as you can follow the tenets of stoicism, you, too, can develop your grit—and resilience—over time. Challenge yourself to continue when the going gets tough. Lean into your self-care and mindfulness practices to keep on the path to your goals. Furthermore, if you still falter on the way to success, take the time to ask yourself why.

Duckworth's book discusses in depth the importance of resilience but hinges even more on your passion. You can have grit and determination—that's crucial—but resilience also depends on your ability to find and remember the passion that guides your why. Once you have achieved this, you can continue to develop grit and become even more resilient. The grittier you are, the more resilient you will become—and the more you will be able to overcome your challenges and succeed in life.[3]

Developing a resilient mindset will serve one well for the duration of their career. The process of creating a resilient mindset isn't for the faint of heart, but neither is life or the fire service. Resilience creates the best chance of not enduring the profession but thriving

Think Critically, Act Decisively

Aside from stoicism and resilience, critical thinking and decisive action—particularly in the face of imperative decisions—are two important abilities of top firefighters and fire officers in the industry. These are not born but developed, as buildable and perishable skills. They are the cornerstones that lead to exceptional fire service leadership. Together, they embody competent, courageous leadership. Being decisive does not always require critical thinking, and critical thinking is not always about being decisive; however, true talent is the ability to conduct both at the same time.[4]

The emergency response environment often requires decisions that are made with urgency owing to unknowns. The truly talented can think critically and act quickly and objectively on a tight deadline. However, some fall victim to what has been called *analysis paralysis*. This is when people cannot move forward at times because they overanalyze, overthink, or second-guess themselves. By the time the crisis is over, indecision becomes a decision itself—and a disastrous one!

Critical thinking and decisive action are complementary, according to Kahneman's *Thinking, Fast and Slow*. Kahneman's argument is that there is a dichotomy: *System 1* is fast, intuitive, and emotional but *System 2* is slow, deliberate, and objective. Kahneman's book discusses the rational and nonrational motivations or triggers associated with each of these systems and further explains how they relate to one another.[5] This is a must-have book for anyone in the fire service, because it truly shows the processes that support critical thinking and decisive acting.

Critical thinking is a competency that prevents assumptions. Leaders think more deeply about the content of the available information, rather than taking the information at surface value; in addition, they consider procuring further information to gather the whole picture. Critical thinking allows organizational leaders to facilitate the evaluation of complex or even conflicting information through a structured methodology. This entails strategic thinking, enhanced problem-solving, and reducing information to general themes and context.

People reluctant to develop these skills may believe that they will rise to the occasion. Despite this attitude, such people do not usually succeed. This term means that individuals work best under pressure. Still, empirical studies demonstrate that people instead sink to the lowest level of the skill mastered and become complacent, which is the exact enemy of critical thinking and the ability to make intelligent decisions. As Gladwell says, "Practice isn't the thing you do once you're good. It's the thing you do that makes you good."[6]

To develop these skill sets, individuals need to flex the mental muscles that control them, as with any skill or strength that needs to be cultivated. This means exercising both sides of the brain—the analytical left and the logical right—to decide on the best course of action. This entails thinking creatively and critically at the same time. While some people believe this to be dangerous, it is only if the critical thinking and decisive action competencies have not been cultivated.

Indeed, we are often afforded time, and those decision might be made after we make a cup of coffee or after we sleep on it. However, emergency responses are the opposite because the compressed timeline means that people must make urgent, critically important, and impactful decisions immediately. In emergencies, the effect of critical thinking and decisive action often looks like, as General George S. Patton quipped, "a good plan, violently executed now, [which] is better than a perfect plan next week."[7] It is essential not to confuse which environment you are operating in; this is because as fire service professionals, more likely than not you are operating in an emergency setting, where quick critical thinking and decisive action are required.

Critical thinking requires reflective, considerate, contemplative, strategic, tactical, and analytical leaders. Highly developed, necessary thinking skills enable leaders to resist the temptation of a "ready, shoot, aim" approach. The best leaders—whether in the fire service or other industries—ask questions, evaluate every piece of information, and are highly empathetic to those impacted by stakeholders' decisions.

To incorporate critical thinking as decisive action, leaders must become responsive, not *reactive*, or *reactionary*. Reactive behavior is often taken without any planning. In other words, reactive behavior is an immediate or knee-jerk response. Emotions, not logic, drive reactive behavior. The need to act decisively and objectively can be an excuse for reactivity, but that often leads to poor individual and leadership performance, which we cannot afford in the fire service.

To explain this further, effective performance requires *responsiveness*. How does responsiveness differ from reactivity? Responsiveness incorporates critical thinking about long and short-term outcomes in the context of the situation. The contrast between these responsive and reactive behaviors is captured in a quote from Wayne Gretzky, "the Great One," describing to his son the key to success in hockey: "Skate to where the puck is going [responsive], not where it has been [reactive]."[8] Being responsive is a calculated, decisive action.

Acting decisively does not mean making quick or hasty decisions. Instead, acting decisively requires narrowing things down to black-and-white considerations while keeping the problematic gray area in the background. Critical thinking can help individuals to evaluate more significant or more complex issues to make a better decision for all. According to Bentson, "A quick and well-thought through decision backed by logic, gut instinct and taking personal responsibility for whatever, the outcome will be, can boost our professional standing in the cycs of thosc around us."[9] Having one characteristic but not the other—such as being a decisive leader and not a critical thinker, or vice versa—can be disastrous. When the certainty of decisive leadership is not complemented by critical thinking, unfortunate, negative long-term outcomes result. This leadership behavior will force an organization to find a new leader to clean up the mess or mitigate the disaster a decisive leader may have created owing to poor critical thinking.[10]

Developing leaders do not always understand what is best for their organization or, even in an emergency, what is the best decision to make. If critical thinking presents data that contradict the course of action, such leaders might ignore this information. This is not an indictment of their ability but a clear example of the opportunity to grow and become more proficient in a necessary skill set.

For organizational health and the entire team's success, be aware that decisiveness can have unintentional consequences that are a precursor to leadership fatigue. For example, it can create an organization of robots, which would impede the leadership growth and integrity of the entire organization. In addition, critical thinking may take a backseat to decisiveness because it increases personal risk and criticism; as a result, the rank and file learn to keep quiet rather than speaking up in situations that require it. Organizational talent can become diminished in these situations, which is not the desired result.

Both critical thinking and decisive action should be prompted by asking these vital questions:

1. Is this an emergency (Kahneman's System 1) or an administrative decision (System 2)?
2. What are the consequences of this decision?
3. What could be the unintended consequences of this decision?
4. What prompted me to need to decide?
5. What is the desired outcome, and who will be affected?
6. What is the worst outcome that can result from this decision?
7. What alternative actions and critical information would be beneficial?
8. Who is lobbying for and who is lobbying against this decision?
9. Is deciding not to decide—or delaying—a viable option?
10. Finally, are you aware of your biases, interests, and intellect regarding the ability to make the decision?

Critical thinking and decisive action require humility rather than hubris. Engaging in critical thinking regularly allows leaders to contemplate their arguments and evaluate what is necessary based on all the available information. Questioning assumptions makes it possible to reconsider the position based on evolving information. Humility also allows decisions to be made with curiosity and analytical discovery. The fire service has some of the most potent leadership of any profession in the world. Thinking critically and acting decisively embody the fire service's leadership ethos and ability to stand confidently alone. Through their application, fire service leaders can build the courage to make tough decisions and the compassion to listen to the needs of others, which are essential in an organization based on trust and strategy.

Chapter 14 Review Questions

1. Why is it important for individuals to exercise both the analytical and logical sides of the brain when making decisions?
   a. To prioritize emotions over rational thinking
   b. To avoid thinking creatively and critically simultaneously
   c. To cultivate the skills of critical thinking and decisive action
   d. To rely solely on intuition in decision-making

2. What does the mental framework of stoicism encourage firefighters to see?
   a. Misfortune only
   b. Circumstances of opportunity
   c. Challenges as insurmountable
   d. Control over external factors

3. How does engaging in critical thinking regularly benefit leaders?
   a. Maintains status quo
   b. Prevents reconsideration of positions
   c. Limits the evaluation of necessary information
   d. Encourages contemplation of arguments and available information

4. Why is questioning one's assumptions important in critical thinking?
   a. To reinforce existing biases
   b. To limit the consideration of evolving information
   c. To make quick decisions without reflection
   d. To reconsider positions based on new information

5. Which of the following are two essential characteristics of top firefighters and fire officers?
   a. Creative thinking and slow action
   b. Critical thinking and decisive action
   c. Slow thinking and reactive action
   d. Rational thinking and emotional action

6. Responsive behavior in decision-making involves considering both short- and long-term outcomes.
   a. True
   b. False

7. Resilience can be developed over time rather than being an innate trait.
   a. True
   b. False

8. Critical thinking is about being emotional and impulsive in decision-making.
   a. True
   b. False

9. Critical thinking and decisive action should be fueled by asking important questions.
   a. True
   b. False

10. Engaging in critical thinking regularly allows leaders to reconsider their positions based on evolving information.
    a. True
    b. False

11. How does stoicism help firefighters develop situational awareness?

12. What does Seneca's statement "Excellence withers without an adversary" mean in the context of stoicism and firefighting?

13. How does having resilience and stoicism benefit firefighters professionally?

14. What is the difference between being responsive and reactive (reactionary) in decision-making?

15. Why is it important for officers in the fire service to learn the stoic principle of wisdom?

15 THE IMPORTANCE OF STRATEGY AND TRUST

Strategy and trust are two of the most critical aspects of any fire service organization. This is mainly because of the vital importance of working alongside team members and maximizing efficiency and proficiency in the field to best serve the community. Plans do not just involve planning, but rather involve a shift in your mindset—in other words, a necessary reframing of the situation to achieve a goal in a manner that you had not tried before.[1]

In the term *strategic planning*, the two words have a forced association.[2] Let's instead evaluate these word individually—and add a related one:

- *Strategy*. We find creative solutions to organizational challenges because of this process, but no methodology or algorithm exists for developing a strategy. It's a creative synthesis process that doesn't come from analysis. Relatively speaking, synthesis and analysis are incompatible.
- *Planning*. This refers to reducing the hole into constituted parts, where strategy is the compound that fuses these ideas.
- *Thinking*. Adding thinking to the equation yields something even more fascinating. Whereas strategic planning works on the skin and bones, strategic thinking works on the organizational soul.[3]

As tools, strategic planning and strategic thinking can be used to realize organization goals—and the goals of the fire service in general—but operate on different levels. In the rapidly changing environment of the fire service, there is uncertainty and a severe need to innovate. Hence, strategic planning alone is not enough to move the fire service organizations.

That is why fire service organizations must use strategic thinking to realize goals and achieve more beneficial results: It is a more thoughtful approach. Several fundamental elements distinguish strategic planning from strategic thinking, and a specific attitude of strategic thinking should be adopted in the fire service. With dwindling resources, increased service demands, and multigenerational workforces, the fire service's strategy without doubt must be augmented. Both approaches, strategic planning and strategic thinking, should be used by senior management and leadership to ensure that fire service organizations operate more effectively. Nonetheless, these are not identical, so the next sections first explore the dissimilarities of strategic planning and strategic thinking before looking at how to shift to strategic thinking itself.

Strategic Planning

Although many people consider this a dated concept, an understanding of strategic planning is necessary before we can go beyond it. Strategic planning is essentially the blueprint for how and what actions should be taken. This is a linear approach, with an upward perspective. Generally, strategic planning focuses on economic, policy, and structural changes within the fire service agency.

Strategic planning requires organizational management to make changes congruent with goals and intended outcomes. These changes could be appointing a new staff, advancing team members, or even moving senior leadership out.[4] The adjustments are typically defined as the *how* and the *what* now approaches.[5] Strategic planning addresses the technical aspects necessary to resolve a problem or improve a process within any organization.

Chaos can typically be mitigated in this process by using an approach that allows for the prediction, analysis, and programming.[6] However, the strategic goals in this approach focus on realistic plans that can be monitored and assessed. However, as Fairholm has stated, strategic planning is useful but limited. This approach is more focused on organizational effectiveness, but there are still gaps to be filled; these gaps are where strategic thinking is necessary. One of the most extensive critiques of strategic planning is that the outcome may not be encouraging. Still, with an understanding of strategic planning, individuals can better understand strategic thinking and its importance as a leader.

Strategic Thinking

In comparison, strategic thinking is more fluid than linear, with a downward perspective. This approach provokes decision-making that is original, research, and even insightful. The context or understanding of complex underlying issues must be known in the face of political, technological, or economic outcomes for any fire service organization. This is the driving force behind any round of changes within the department, and the vision will then be aligned with the goals to remain competitive by developing unique and innovative strategies.

Heath suggests that by applying upstream thinking—that is, imagining and preparing for the future ahead of time—we can prevent problems from occurring in the first place.[7] These strategies will likely acknowledge the internal and external factors influencing the fire service agency and the future strategic planning that they would like to pursue. Therefore, strategic thinking takes precedence over strategic planning.

The Focus on Thinking Strategically

In our highly chaotic world today, leaders must think strategically to overcome the challenges they are presented with. This type of leadership is grounded in agency core values, which are supported by the mission and vision statements of the entire department. Strategic thinking allows a solution to be reached through brainstorming, which often leads to further success. By contrast, strategic planning is more centered on managing projects or tasks.

Kim and Mauborgne coined the term *red ocean* to describe fierce competition in the business world.[8] This analogy is relevant to the fire service because a few external and internal factors affect any fire service organization's success. Finding an agile approach to control the chaos requires focusing on the organizational values and their purpose; it can even lead to further success.

Strategic planning is focused on achieving the desired goals, but strategic thinking takes a holistic and philosophical approach that is undoubtedly significant for the fire service agency. Its purpose or focus is much more on responsibility, and it is far more adaptable and responsive to changes that allow the fire department to stay focused on its goals while not losing any of its progress.

The Importance of Information Sharing

In addition to values, information sharing is an essential aspect of organizational strategy. Counterintuitively, despite the accessibility of information in the internet age, sharing timely, relevant, and appropriate information remains a challenge. Strategic planning might have controlled or limited access to information, because relationship building and feedback are essential in creating trust within an organization. However, information sharing can make the difference between preventing a problem with proper upstream planning or failing to do so, leading to more significant challenges later.[9]

Information sharing is, without a doubt, a paramount aspect of organizational strategic development. The trust that can be developed from these relationships will allow the leader and their intent to be received positively, which will further solidify the agency's goals. This means that building a culture of trust and information sharing is more effective than building an organizational structure, which is the goal of strategic planning. This allows the organization to solve its internal issues, build itself up to resist threats and challenges, and prepare for the future.

Change in Focus Must Embrace Uncertainty

In addition, fire service agencies must change to embrace uncertainty throughout any situation. Strategic thinking is a reply to what one should be doing and, more so, an approach aligned with innovation and a pathway to success.[10] However, the focus is on people rather than systems.

For example, Google has applied strategic thinking by allowing engineers to devote a portion of their company time to side projects, through which their employees cultivate new ideas that lead to innovation of potential new products. Embracing uncertainty within the fire service can also have a similar effect because it will allow for effective and efficient adaptation and innovation. This puts the agency in a healthier position to offer the community more value. This point is supported by a study by Kazmi and Naaranoja that found a strong linkage in the business world between product innovation and strategic thinking, which the researchers termed *target reach*.[11]

Forward Thinking and Planning for Strategic Success

Forward thinking and planning is the proactive approach of anticipating future trends, challenges, and opportunities, and preparing for them in advance. It involves a strategic mindset focused on long-term goals, innovation, and adaptability, allowing individuals or organizations to stay ahead of changes and capitalize on potential developments. Forward thinking goes beyond immediate concerns, requiring an understanding of emerging technologies, shifting market demands, or evolving social needs. Planning complements this by setting a structured path toward these anticipated objectives, creating actionable steps that prepare for various outcomes and help mitigate risks. Together, forward thinking and planning enable resilience, sustained growth, and an enhanced ability to respond to unforeseen circumstances.

Fire service agencies using strategic thinking as a core approach will include innovation for forward thinking and planning in their organizations. This has an immediate effect on a fire department's success. These organizations will become more prone to understanding how to respond to external and internal threats.

Through strategic thinking, departments will follow a more customer-oriented approach while remaining eager to develop new strategies that allow for the further development of organizational culture. Strategic planning and thinking are codependent, and offer shared quantitative and qualitative value exchanges. When strategic thinking is omitted, an organization will not be able to effectively address these changes to the environment, whether internal or external. This could further cause the organization to fail to innovate.

Fairholm has asserted that strategic thinking will "link the organizational soul to the body that is rightly fit by organizational managers and planners."[12] The "organizational soul" embodies the collective spiritual values, ethical standards, and sense of community within a workplace, reflecting the deeper purpose and meaning that guide its members. This concept emphasizes fostering wholeness, integrity, stewardship, and morality among stakeholders, encouraging leaders to focus on the spiritual well-being of both themselves and their teams.

Therefore, strategic planning and thinking are essential because they will not only unlock the innovation within your organization but also link this organizational soul to the planners and managers, making yours a more efficient team for you to guide to success.

Understanding the Importance of Trust

Beyond the various benefits of strategic planning and strategic thinking, trust is required of organizations and their leaders. Trust is the willingness to be exposed to the actions of another person or group; it is the belief that both individuals act beneficially and that they understand the true intentions of everyone.

Leadership and trust are essential elements that represent the best versions of our fire service culture. Various movies, including *Backdraft*, have highlighted the fundamental importance of trust within the fire service organization. The trust and leadership required in the fire service must be unbreakable, for mutual, coordinated action, shared values, and a shared vision. If you couple this with strategic planning and thinking, this creates an environment where every person can trust their purposes and their actions and further align with the intention of the organization for tremendous success and innovation.[13]

Building a Culture of Trust

We have talked a lot about organizational culture in previous chapters, but I want to impress upon you the importance of building a balanced, collaborative fire service team. It is not just about curating the favor of each team member but about leading and creating this for the team. This entails understanding the common goals and principles of the organizational culture (as discussed in chapter 2). This is a process and one that encourages trust over time.

Sadly, not many organizations have sound cultures, and leadership is often not as dynamic unless it is shared and compatible with culture and vision alongside that of values. Leadership requires individuals in these top positions to go beyond themselves, embracing tenets of servant leadership and becoming technically competent to foster a vision in a set of values that creates an unbreakable culture.

In addition to benefiting the organizational culture, healthy relationships (in any capacity) are built on a foundation of trust. Effective communication, teamwork, employee commitment, and productivity rely on trust. All these aspects together create the perfect positive department culture that any fire service professional should aim for.

Trusting relationships are meant to be inherently vulnerable, so trust is undoubtedly earned. A lack of trust is often seen between coworkers, managers, and even employees in the fire station. Furthermore, trust is a dynamic concept, so there are several distinct types of trust.

The Various Types of Trust

Once you understand the different dynamics and types of trust, you can understand how to foster this within your organization, which could lead to further success in your strategic planning and thinking. Plenty of research has been dedicated to trust in the workplace, and models allow us to see how trust is built and maintained. One worth mentioning is Michelle and Dennis Reina's model of trust and betrayal, which outlines three critical components:[14]

- *Capability trust* is a type of trust that develops when people understand another individual's perceived knowledge, skills, or abilities.
- *Character trust* is the belief that individuals will do what they say they will do and that this can be trusted, rather than second-guessing what somebody has said.
- *Communication trust* is the willingness to share information, to tell the truth, to admit one's mistakes, and to keep secrets.

Forces That Hinder the Development of Trust

In addition to the various types of trust, it's essential to understand the forces that can hinder trust in organizations. If you want to have a successful organization, as a fire service leader, you must take on the risk and difficulty of essential tasks; you must also understand and address any barriers to harmony and trust that exist. Otherwise, your organization will fail to be successful.

The immovable forces around the individual and organization can often be roadblocks unless handled appropriately. There are definitive solutions, but often, they are not readily available to us. Therefore, the takeaway message is that recognizing these challenges and minimizing their impact requires creativity and finesse, skills that ultimately increase trust and allow leaders to succeed regardless.

Individual Forces

Our organizational members communicate information about feelings, attitudes, problems, and ideas whenever they interact with anyone else in the workplace. Such individual preconceptions that frame communications in the workplace carry danger because of the acceptance of information that we deem to be either accurate or helpful and the perception we may have about whether they the source(s) are or are not trustworthy.[15] The converse is also true: Erroneous or deceptive communications can erode trust. This is played out around the kitchen table in every firehouse in the country.

According to Culbert and McDonough, other human tendencies related to communication get in the way of trust.[16] People don't like subjectivity because it's so difficult to control. However, most verbal communication is subjective and not supported by objective evidence. Administrators constantly face challenges with trust, and in the absence of critical information sharing, trust will face even greater jeopardy.

Different people could interpret the same event in many ways, as perspective changes absolutely everything. Leaders should not make the mistake of assuming that others in the organization share the same perspective they have, because every individual will have a different perception of what has happened and what the facts truly appear to be. Therefore, every individual deserves the agency to believe what is and is not true for themselves, to a degree—in other words, their point of view.

Leaders must allow others to be empowered; nonetheless facts must always be preserved, and the consensus truth must always come first. Some leaders are unwilling to enable trust and reluctant to harness the power of their people by empowering them. Power should not be focused upon the leader or follower; instead, a delicate balance is imperative in any organization so that all members may take part in the critical decisions and allocation of resources. When a leader or follower disproportionately concentrates power on themselves, this creates a barrier to trust.[17] Power abuse undermines trust.

People have mistaken trust for belief in the goodness of others and in the world having a benign nature. Some consider those who trust themselves to be less intelligent and more gullible; they believe that having trust issues lowers a person's self-worth or potential for success, so they label them as potentially naive or incapable of completing a task. This cynical mindset can stifle the ability of the leaders to succeed, making it more difficult for the team as well.

Trust will consistently be earned over time, because of the individual's life experiences. Those who can trust have more experience with trustworthy others, and trust is a belief in communication, interaction, and relationships—and their associated honesty—rather than in correctness. In other words, having a high level of confidence in a person is about being able to rely on them to be trustworthy and honest. Trust is risky, but so is the mantle of leadership, which should be seen as an opportunity to be vulnerable and understand and mitigate the risks of trusting others. To succeed, leaders must find a way to understand, trust, and reduce the risks of instilling trust in every team member.

Note, however, that people have self-serving motives. These are healthy and normal but still need to be recognized and understood. Members of the organization can compete with others in social relationships, including work, to achieve their goals. This competition can create a win-lose scenario, focusing on individuals rather than group action, which makes high-trust relationships extremely difficult. Self-interest exists, but it directly opposes trust, because the values that support high trust are not the same that help an individual or even your self-interest.

Organizational Forces

The second challenge to developing trust comes from organizational forces. People's actions and positions of authority are critical to developing or destroying a trust culture, typically because the cumulative action shapes the organization's internal climate in many ways. Leaders may use their power to promote or support a culture that values high trust. However, in other cases, they may use it to create distrustful atmospheres, even without intending to.

The high level of authority instilled in an organization's activity versus the level of self-direction allowed in a relationship between leader and follower is the key to what kind of culture is ultimately developed. Leaders have complex roles in a complex world, so there should be structured, institutional authority to promote individual growth and group success.

Societal values influence fire department values, a trend often associated with deconstructing traditional values and morals. Trust is linked to these, which leads to the decline of fundamental values like honesty, integrity, dependability, and even commitment. This may make a leadership's effort to create a trust culture even more difficult. If leaders do not face this ethical deficit, fire service organizations—and even the country—will end in shambles.[18]

These roadblocks to trust can be incredibly powerful. They are specific to individuals or even departments, with unintended negative consequences that can even sometimes be described as pathological. Mistrust reduces group or individual success.

In addition to the individual and organizational forces that may obstruct the development of trust, more general institutional and human barriers to a leader's ability to instill trust in an organization exist. These include traditionalism, uncontrolled growth, politics in the office, or even cynicism or burnout. While the magnitude and effects of these barriers vary depending on the organization, they appear to be present in all groups to a degree.

Ultimately, understanding the roadblocks to trust within the organization can create an even greater understanding of how to develop trust. Additionally, it's important to recognize the challenges of overcoming self-interest versus the necessity of team orientation, and how this affects building trust.

The Power of Strategic Planning, Strategic Thinking, and Trust Together

In any organization, the power of strategic planning, strategic thinking, and trust can ultimately change the nature of any fire station. In the fire service, trust is finally the building block to ensure that critical thinking and strategic planning are eventually instituted and applied in the best possible manner. Without a culture of trust, strategic planning and thinking become less effective, which is why it is so critical this is discussed in the fire service.

Regardless of leadership philosophy, how you build your culture, or how you interact with your employees, power comes down to one factor alone, and that is trust. Without trust and strategy, your team and your organization will not have the same impact. Therefore, I urge you to identify ways to improve the trust within your organization, so that you can move forward effectively for the maximum benefit of the organization and community.

Chapter 15 Review Questions

1. Which of the following best describes self-serving motives in individuals?
   a. Uncommon and should be discouraged
   b. Unhealthy and abnormal
   c. Directly align with building trust
   d. Normal and healthy but need to be recognized and understood

2. In the fire service, why is trust considered the building block for critical thinking and strategic planning?
    a. Trust has no impact on critical thinking and strategic planning.
    b. Critical thinking and strategic planning can function effectively without trust.
    c. Trust enables critical thinking and strategic planning to be implemented effectively.
    d. Trust hinders the effectiveness of critical thinking and strategic planning.
    e. Trust is unrelated to critical thinking and strategic planning.

3. How should strategic thinking and strategic planning be prioritized in a fire service organization?
    a. They are interchangeable terms with the same meaning.
    b. Strategic planning should always take precedence over strategic thinking.
    c. Strategic thinking is more crucial in today's rapidly changing environment.
    d. Strategic planning alone is sufficient to address organizational challenges.

4. Why is information sharing crucial for organizational strategic development in the fire service?
    a. Information sharing can limit relationship building within the organization.
    b. Information sharing may lead to increased uncertainties in a fire agency.
    c. Building a culture of trust can inhibit information sharing.
    d. Information sharing allows effective responses to internal challenges.
    e. Trust developed from information sharing can confirm the agency's goals.

5. Which of the following is a best practice for overcoming the challenges of self-interest versus team orientation?
    a. Ignoring challenges
    b. Prioritizing self-interest over team orientation
    c. Recognizing and understanding self-serving motives
    d. Disregarding trust altogether

6. Strategic planning alone is sufficient for organizations to address environmental changes effectively.
   a. True
   b. False

7. The presence of trust is crucial for ensuring that critical thinking and strategic planning are optimally implemented in any organization, particularly in the fire service.
   a. True
   b. False

8. Leadership and trust are not crucial elements for the success of a fire service culture.
   a. True
   b. False

9. Communication trust involves the willingness to share information, tell the truth, admit mistakes, and keep secrets.
   a. True
   b. False

10. Strategic planning and thinking in a fire service agency are completely independent of each other.
    a. True
    b. False

11. How can understanding the concept of trust dynamics help fire service professionals foster a culture of trust within their organizations?

12. How do traditionalism and cynicism act with regard to instilling trust within a team?

13. Why is trust considered a building block for effective strategic planning and strategic thinking?

14. How can fire service leaders integrate strategic thinking into their decision-making processes effectively?

15. How can fire service organizations balance the need for strategic planning and strategic thinking to adapt to evolving environments?

16 LOVE, GRACE, AND TEMPERANCE IN THE FIRE SERVICE

Trust in the fire service entails much more than just having the best strategies, being a great leader, and even understanding the fire service itself. Ultimately, encouraging trust and a great team atmosphere comes down to three essential concepts: love, grace, and temperance. People often forget these, assuming that leadership is mainly deduced from theories, methodologies, and perspectives. However, if you cannot remain compassionate, caring, graceful, and balanced in good and bad situations, your leadership and your fire department will suffer.

If the fire service is about stoicism, resilience, and accountability, then why love, grace, and temperance? Without love, grace, and temperance, the fire service itself struggles to make the connection that it needs with the community. Even more important, these three are principles underpinning effective fire service leadership. They are absolutely essential to create an environment predicated on compassion, respect, and trust. This final chapter holds a message for all of us to embrace as we embark on our leadership journey in the fire service.

In this chapter, we will discuss how leadership in the fire service is rooted through creating a culture where firefighters feel valued, supported, and empowered to make decisions and fulfill their duties. We will talk about the power of grace and how recognizing the efforts of others and even their failures is crucial to motivating, inspiring, and overcoming significant adversity. Finally, we will explore the importance of balanced leadership through temperance by setting the tone, exercising self-discipline, and embracing moderation as an absolute minimum expectation within your fire department. With these three concepts, leadership in the fire service goes from great to extraordinary and can even be awe inspiring.

How Love Fits into Fire Service Leadership

Love in the fire service has as much of a place as stoicism. Even though many of us want to resort to stoicism and the resilient face we put on for the world, love does have a place, and it is essential that you, as a leader, recognize this. Unlike love in other systems or organizations, love in the fire service is often most visible when looking at how leaders inspire and motivate their fellow firefighters. Love encompasses—in its most fundamental definition—care, compassion, and support, which are all seen within the fire service, even today.

Caring for others in the fire service is an essential part of being a leader. Without caring for your fellow firefighters, it is difficult to get just about anything done or to encourage any form of trust and camaraderie. In many leadership positions, caring is as simple as asking employees how they are and seeing how their days are going. Often people find it reassuring when someone is thinking about them or checking in on them to see if they are handling their responsibilities effectively.

All firefighters experience moments of weakness—even those who resort to stoicism and avoid confronting many challenging moments they endure on their job. This is when love shines through. As leaders, it is our job to recognize when a firefighter needs to be picked up when they are struggling and when a firefighter needs to be seen and recognized for the great sacrifices they have made for the community.

Despite any preconceptions to the contrary, love does have a place within fire service leadership. It can be applied to almost every aspect of fire department ethos, for both the highly successful events and the most harrowing moments on the job. Leaders comfort their fellow firefighters when troubling events occur, and leaders congratulate their fellow firefighters when the job is exceedingly well done. Integrating care and compassion into the way that you lead will not only yield significant benefits in encouraging a more positive organizational culture but also build relationships and trust within the fire service, often at an unprecedented level. Give your organization the benefit of experiencing love.

How Love Builds Relationships and Trust Within the Fire Service

Being a stoic leader and somebody who can put on a game face when times are tough is essential for the moments in which your fire department needs strength to lean into. However, to build strong relationships and encourage trust within your fire department, you must lead with and promote care, compassion, and empathy within your organization.

Without care, compassion, and empathy, it will be extremely difficult to build relationships with your fellow firefighters. Beyond establishing organizational culture and building trust, a successful organization ultimately comes down to how you lead—and especially how caring and compassionate your fire department's leadership is to the challenges your members face every single day. Love can also lead to more compassionate communication.

For instance, if your fire department has just endured a significantly harrowing event on the job—from which many of your fire department are still recovering, often more mentally than physically—this is an ideal opportunity for you to exercise compassionate communication. Leave your door open, literally and figuratively, for your firefighters to come and discuss any challenges or issues they feel arose during or since the event. Hold a meeting and debrief them, with compassion and empathy toward your fire department. Be vulnerable and honest about how things happened and even how it affected you. It may feel like the wrong time and place, but firefighters need to see that even leaders have their moments and are affected by challenging jobs.

You don't need to be exceedingly emotional to demonstrate that you have a caring heart. All you need to insert compassion into your leadership is simply to allow your fire department to recognize that you understand how they feel. Be honest about your emotions when asked. Also ensure that compassionate communications are relevant; essentially, all the firefighters need is the confirmation that they are not alone and that the feelings they are experiencing are expected, rather than abnormal.

If you are worried about encouraging more compassionate communication, there are many methods to create a team atmosphere resting on empathy and compassion. For instance, hold regular department meetings and invite the membership to add items to the agenda, whether bringing to light concerns or discussing the general comfort of their fire department. Furthermore, having an open-door policy is a great way for a leader to encourage trust and build relationships while also ensuring that your firefighters feel valued, supported, and heard, no matter what they may be dealing with.

Building trust by engaging in open, compassionate communication is key to forming relationships based on the compassion that you have for your fire department. When you display that you genuinely care for every member of your fire department, you will see extraordinary events happen because of the relationships that you build, the camaraderie that is developed, and even more importantly, the trust and care that each firefighter has not only for themselves but for every individual who they work with on the job. However, conflicts are inevitable, but how you lead with compassion will also resolve these challenging moments.

How Love and Compassion Resolve Conflicts

While integrating love and compassion into your leadership style can ultimately build relationships and trust, love and compassion are also necessary in the face of inevitable conflicts. In any organization, there will always be challenges and conflicts that arise. It is not only your job as a fire service leader both to foresee these when they happen and to mitigate them appropriately. Remaining stoic, resilient, and calm is crucial when addressing any issue. Still, you must also be compassionate because stoicism and analytics do not resolve all conflicts.

Resolving conflicts with love and understanding is essential in the fire service for several reasons. The fire service is arguably one of the most challenging of all professions. Having been in the industry for many years, I have seen several examples of the strength firefighters carry daily when they choose to carry the fire and lead. They remain resilient for their loved ones, community, and themselves.

The first reason that love and understanding are absolutely essential when resolving conflicts in the fire service is that they help us to understand that others have vastly different experiences in life. Growing up, many of us were told, "You have no idea what other people are going through." Therefore, whether you're a leader or not, showing compassion is essential to resolving conflicts peacefully and respectfully.

The second reason that resolving conflicts with love and understanding is absolutely essential in the fire service is because it will ultimately bring about the opportunity for mutually beneficial solutions. The ultimate goal of any conflict resolution should be to find a mutually beneficial situation for both parties involved. For example, if there is a disagreement within the fire department, it is helpful and respectful to try and find a way that both parties will benefit. For example, as a leader, you might mitigate a situation where your fire department finds an appropriate compromise to use the resources available when they are within the fire department for their shift.

Resolving conflicts with love and understanding must be an expected practice regardless of your environment. In the fire service, as the leader of your organization, it is paramount that you uphold these concepts as integral to the way that you lead, even in the most challenging times.

To navigate challenging situations, consider how the other person feels before you do anything else. As a leader, your job is to understand your entire team—how they are feeling, what they think is wrong about a situation, or what they feel should be done. By understanding what both sides of the conflict want from a situation, you can devise an appropriate solution that benefits all parties. Think of yourself as the mediator for any conflict resolution. With a compassionate, caring perspective as the leader, you can find a reason for both parties to compromise, and that is ultimately the goal.

Your agency will always encounter challenges, no matter how strong or resilient it may be. This is inevitable because, as people, we constantly find ways that or reasons why things are not as they should be. With love and support for every member of your fire department, you can lead the way, avoid and mitigate conflicts, and create relationships that are truly built on trust and understanding, which should be the cornerstone of any fire service organization.

Grace and Leadership in the Fire Service

Aside from leading with compassion and understanding, you also need grace in your leadership style. Not many in the fire service associate grace with how they lead, but it is an essential aspect of how you lead and relate to others.

Having grace means understanding, forgiving others, and being willing to extend compassion, even in situations you otherwise wouldn't. Being graceful is not about avoiding failure, but rather carrying yourself in a manner where people respect you and look up to you as a voice of reason and support. As the leader of your organization, this should be your ultimate goal because you should be the mediator, the voice of reason, and a source of strength and inspiration. You shouldn't expect yourself to be perfect, because then your team would expect the same of themselves—and that simply isn't possible for anyone.

How to Cultivate a Culture of Forgiveness and Understanding

A culture of forgiveness and understanding is crucial not only for you as the leader but also for the safety and comfort of your fire department. The importance of forgiveness, especially in such an industry as the fire service, is often understated. Being forgiving means recognizing that your fire department will make mistakes just as much as you will. Moreover, it acknowledges that there's a time and a place for mistakes. Ensuring that your fire department understands this and accepts their mistakes will be essential for their mental health.

As leader of the fire department, you must promote professionalism, ethics, and morals as the foundation for every decision. In some situations, forgiveness will not be easy—or even appropriate. When breaches of ethical conduct or unacceptable behaviors risk the safety and well-being of others, the appropriateness of forgiveness should be weighed carefully. However, for most other situations, forgiveness should be an expected practice. This allows you to build stronger relationships with your fire department and stand out as a leader who understands and is compassionate toward the challenges others face.

For example, when training, fire department personnel can make as many mistakes as they would like simply because it is their time to learn how to make these critical decisions and work with their fellow firefighters. In a culture of forgiveness and understanding, this is the foundation. Training gives opportunities for firefighters to accept their mistakes and apologize for making a blunder that would be significant in the field.

These opportunities allow firefighters to not only be themselves but grow—to become better and to forgive themselves when they make mistakes. Thus, each mistake made is also an opportunity to forgive, but not forget, and move forward. This enables teams to become even more successful by being more prepared for each emergency that your fire department faces in the field.

Cultivating this culture from the ground up—from when you first trained your fire department to when they are in the field—is paramount. Train your firefighters in the training environment to apologize when they make mistakes, to recognize what they did, and to expect that you understand what they did and how they will improve. Through this, they grow, and more important, it cultivates a safe atmosphere for them to become the fire department they are destined to be.

Each and every firefighter has an opportunity to grow into a fantastic leader. Whether you are a leader yourself now or aspire to be a fire department leader, forgiveness and understanding will allow you to become the leader you were meant to be and the leader you desire to be. Without these at the base of everything your fire department does, it is nearly impossible for firefighters to become influential leaders.

As the leader, there are also ways that you can nurture forgiveness and understanding in yourself. When training in the field, make yourself approachable. Do not miss opportunities for education. If a firefighter makes a mistake, do not scold them, but instead let them come forward in conversation; that way, you can forgive them for the mistake and coach them on how to do better. This dynamic not only will position you as the leader that your fire department needs but will also make your personnel more comfortable coming to you, as well as going to other firefighters for assistance. This is especially relevant in the act of firefighting when a real emergency hits.

How to Demonstrate Grace in Your Leadership

As a leader, you can demonstrate what it means to lead with grace by exhibiting patience and compassion. In such a high-stress environment as the fire service, being a leader who exudes grace shows your fire department how to follow, regardless of whether they become chief. This section outlines a few strategies that I recommend.

One method that demonstrates grace is admitting when you were wrong. Grace means knowing when you've made a mistake and recognizing that it needs to be addressed. For example, say that you made an error or an inaccurate comment during training. As the leader, this is your chance to admit that you made a mistake, to address it, and then to guide your team in the correct direction. Firefighters will respect you even more when you acknowledge your mistakes. After all, nobody wants to follow a leader who thinks they are perfect.

Another method of leading with grace is showing compassion through your actions and decisions. There's a time and a place for being stoic; there's also a time and a place for allowing your emotions to show, even if just for a moment. Show compassion for your fellow firefighters, recognize when firefighters are struggling, and offer an opportunity to connect with them, listen to them, and be there for them. Grace means showing compassion, seeking understanding, and having patience to address together the unique challenges that firefighters face.

This is even more important if you have many new firefighters in your fire department. Allowing them to adjust to the high-stress environment of the fire service will benefit them in the long run. Furthermore, it will encourage them as firefighters to follow in the steps of their leadership, listen to the fire service veterans, and see that their challenges are not theirs alone. With these tips, you can inspire your fire department to reach higher for even more admirable goals, including, importantly, personal growth and learning.

How Graceful Leadership Encourages Growth and Learning

For an organization where growth and learning are expected daily, encourage grace through how you lead. Be compassionate, be understanding, and most importantly, allow for forgiveness in the moments that matter. Graceful leadership ultimately cultivates a learning environment from which each firefighter benefits, where continuous improvement is the expectation. Your firefighters will follow you and succeed, benefiting from your support of their development and growth.

When you train and empower your fire department to experience growth and learning, you are creating better firefighters. Moreover, you are cultivating even stronger men and women who you serve with every single day as the leader of the fire department. Your fire department will become even more robust when emergencies arise, and your personnel will demonstrating themselves as the successful, critically prepared, and stoic individuals you know them to be.

The Role of Temperance in the Fire Service

Closely related to stoicism is the concept of temperance. Temperance is related to moderation and self-discipline, which are topics we have already discussed (in particular, see chapters 6 and 13). Many people often do not associate this with the fire service, and you may wonder why this topic was saved for last: leaders must understand self-control, self-discipline, and accountability first.

Through self-discipline, you ensure that your leadership is as sound as possible for your fire department. After all, you would lead your firefighters down a difficult path if you could not moderate your own actions and decisions. In addition, your self-discipline—in the form of restraint—creates space for thoughtful decision-making that impacts the entire organization and your fire department, as well as the community beyond the fire department. It is not without the leader's restraint that the fire department succeeds during certain situations.

At times you will want to call out firefighters for acts that led to mistakes. There will even be times when you feel you have no other choice than to be upset. Nonetheless, you must lead your fire department with temperance despite the challenges you may be faced with individually. Remember, the fire service is one of the most high-stress environments you can encounter, and as a leader in this industry, you will likely face more stress than anyone in your fire department. Firefighters are the strongest of the strong, and that's why temperance holds a significant place in fire service leadership.

Temperance as Personal Accountability

Temperance is a form of personal accountability. Incorporating temperance into your leadership ethos allows you to moderate your actions, to lead responsibly. Thus, professionalism, ethics, and expectations will guide your fire department down the right path. Your fire department needs a strong and reliable leader, and with temperance, you meet that demand and even set an example for them as they aspire to become leaders in the fire service. Furthermore, your accountability is as beneficial for you as it is for your fire department.

As I can attest firsthand, there will be many throughout your journey in fire service leadership when it will be increasingly difficult for you to meet certain challenges. No leader is capable of handling all that is thrown at them. We are pressured to become the leaders who we become because of the high stress that we mitigate as commanders of our fire departments.

How Temperance Aids in Managing Emotions and Stress

As mentioned in chapter 14, handling situations with stoicism is crucial at times, and temperance undoubtedly has application at these same times. For instance, temperance helps a leader to determine whether to use compassion versus stoicism.

After a challenging emergency response, for example, it can be difficult to mentally process what has happened. This is natural for anybody in the fire service (or any high-stress environment). In these moments, take a chance to collect yourself, process what happened, and then go to your firefighters. Importantly, do not dwell on your own emotions because others rely on you for guidance. Remember: your fire department relies on you as their leader to guide them to move past challenges, too. Thus, you can move through challenging moments in tandem with your organization.

When you debrief your firefighters, exercise temperance. Be stoic but compassionate. When in doubt, always be the leader you would have needed in their position, as a firefighter. Ask yourself how you would want your fire chief to act if you were sitting in the crowd listening to the debriefing session discussing what just happened? By reframing your perspective and consider your firefighters, you will never make the wrong choice for your organization or for yourself. Then, you will find yourself acting with temperance, managing the stress and the challenging emotions that come with being a leader in the fire service.

Temperance is beneficial for moments in which you need to manage the emotions and stress within your fire department while also being aware of and managing your own emotions and stress. Temperance allows you to recognize when it's time to take off the stoic mask and truly process what happened, even if more emotionally than desired. In the fire service, there will be times that test you and times that surprise you. That's how the industry works.

Sometimes, you may even wonder why you committed yourself to this profession. The fire service is not for the weak—that is something we all recognize. You chose the fire service because you are a strong individual and wanted to be a pillar for your community. Doing so requires you to be strong, often more so than you may feel inside. With temperance, you allow yourself to grow into a strong, compassionate, yet still stoic individual.

A potential emotional reaction to something that happened, whether personal or professional, can be not only a challenge but also a learning opportunity for you to improve your decision-making and the teamwork of your entire department. If you act too emotionally, impulsively, or without reason, this can be highly damaging to your organization, which relies on you to make vital decisions.

Take the time to react emotionally when needed, but do so in private so that you can process first, then come forward to speak to your fire department with the right frame of mind. Your firefighters need somebody solid and compassionate, but it is acceptable and normal for you to have moments when you are at an emotional weak point. By practicing stress management and emotional regulation, you can truly overcome these moments by working through them appropriately to process the emotions you may be dealing with because of the high-stress environment of your job.

How to Make Sound Decisions—Even under Pressure

Once you master the practice of temperance to manage your emotions and stress, it can also be used to aid your decision-making in high-stress environments. Decision-making in high-stress environments is never going to be easy. You must make critical decisions during your time in the fire service, and some are life or death for those involved. In the fire service, a number of factors force you to be in the line of duty and in the face of danger, and in these, true leaders demonstrate themselves.

If you have mastered the art of temperance in your leadership and self-management, then managing high-stress environments and decision-making will be straightforward. It is impossible to make decisions in high-stress environments be easy, but you can manage them even more effectively if you understand when to react and act.

Prioritize effective decision-making. In the face of time-sensitive decisions, as is often the case in the fire service, make sure you understand your priorities. Only once these are clear, then decide as quickly as possible. Every minute counts, and for your fire department, every minute you take to choose is a moment they wait to act.

To improve your own and the decision-making skills of your firefighters, insert many learning opportunities within the training sessions. Hold debriefing sessions where you discuss what could have gone better, where you recognize what may have gone wrong, and where you commend people on the effort that they put forth to make effective decisions—including decisions that ultimately saved people's lives. It is critically important for you to exercise temperance to make decisions, and doing so sets an example for firefighters to recognize how they can make effective decisions for themselves.

After all, your role as the leader is to guide the next generation to be able to take the helm of the fire service—when as the new officers and chiefs they will need to effectively lead and carry on the tradition of the stoic, resilient, compassionate fire service. The more you exercise your opportunity to improve your decision-making and better manage the high-stress environment you encounter in your leadership, the more you prepare your fire department to excel and truly exemplify what it means to be a firefighter.

Integrating Love, Grace, and Temperance

All Three as One Philosophy

In the fire service, some may regard love, grace, and temperance as optional topics. However, these remain ever critical and are intertwined as an essential part of what we expect from effective fire service leadership. Now that we have discussed love, grace, and temperance separately, let's combine them into one all-encompassing philosophy. These are mutually reinforcing concepts that can change how you lead—and even how you live your life. With empathy, grace, and temperance, the way you teach and the relationships you have in your life will change drastically.

It is with compassion that leaders are motivated to become even better than they were the day before. We work as fire service leaders to encourage our firefighters to become even better than they were the day before. I work every single day because of the compassion and the care that I have for every member of my fire department.

Without every firefighter, our fire service organization would not be the same. I am grateful every day for the efforts that they put forth. It is my duty and my responsibility as the fire chief to allow them to see the respect and understanding I have for them, the empathy that I carry, and the understanding of the burden that they have chosen when they became firefighters and entered this industry.

It is with grace that I, as the fire chief, also see change in how I lead and guide my firefighters over time. Firefighters within my fire department rely on a strong leader, and it is my responsibility to always be a leader who exercises grace, humility, and forgiveness. If my fire department feels I am unapproachable, expect perfection, or have never made a mistake, then I have certainly done something wrong as their leader. My philosophy is to lead with compassion, grace, and temperance to set an example; thus, as leaders we become more approachable and make firefighters more coachable.

I have seen significant shifts in how I lead my organization thanks to the inclusion of love, grace, and intemperance. Turning these three tenets into a single teaching philosophy has changed how my firefighters view me, themselves, and each other. Regardless of the stoicism that remains a significant tradition within the fire service, fire leaders should practice compassion, grace, and forgiveness with every firefighter. With each opportunity to forgive mistakes and misunderstandings, we build more vital trust and relationships. Ultimately, this benefits the people who matter the most in this equation: the community. Incorporate all three into your leadership practice.

If you find yourself struggling to envision how a single leadership philosophy can encompass love, grace, and temperance, there are a few strategies I recommend. This section contains a multitude of examples from my own fire service experience, as well as from others that I have seen in the industry. Let's first talk about encouraging more compassionate, forgiving conversations in the face of disagreements and conflicts.

When disagreements and conflicts arise, embrace love, grace, and temperance. For example, in let's consider a disagreement about the way you have chosen to lead the organization, where a firefighter has come forward expressing concern. In this case, first allow the firefighter to explain themselves and practice grace and compassion while trying to understand them before you explain your perspective. Importantly, avoid wanting to react before you appropriately respond; take a moment and reflect on your own emotions, then seek to understand and seek forgiveness. Within these moments, you encourage understanding, exercise noticeable compassion for their perspective, and moderate yourself to be the leader they need, rather than a person responding emotionally to something that does not require such a reaction.

In challenging moments, such as great loss or difficult emergencies, love, grace, and temperance are essential. Throughout my time in the fire service there have been moments that tested my grace, temperance, and even compassion.

Loss tends to stretch the bounds of compassion, forcing us to react emotionally to moments beyond our control. Even if we do everything right in certain situations, we cannot stop something from happening. The loss of human life or material goods is possible—and may even be unavoidable—in grave situations. Within these moments, I urge you to have compassion for yourself and not force yourself to remain stoic after the moment has passed.

Allow yourself to be emotional but exercise temperance. Self-moderation is still necessary because at the end of the day, regardless of whether you are in a leadership position, your fire department relies on you to be as strong as you can be. Within such harrowing moments, even if you choose not to express how you truly feel to your fire department, your colleagues understand, respect, and lean on you and each other to overcome adversity. You can survive the gut-wrenching losses by including love, grace, and temperance. In the fire service, it is paramount that you do. Allow these moments to strengthen you and encourage your firefighters to try even more to improve themselves and overcome the challenges that they encounter.

In the face of significant loss and challenges, it is difficult, even for me, to avoid feeling emotional. However, the right balance of love, grace, and temperance can allow you to moderate your behavior, respond as a leader should, and process what has just happened. A leadership philosophy including all three of these tenets can influence the way that you lead and the way that you relate to your fire department.

By exercising such a leadership philosophy, you model behavior and encourage compassionate communication. Even though you may be able to handle an emergency situation and exercise self-moderation, many firefighters require a compassionate ear to listen, in addition to a helping hand.

As a leader, I urge you to make yourself available. The more understanding and open you are, the more inspirational, supportive, and respected you will be. Most importantly, you will grow your firefighters into the men and women they are destined to become, and you will be part of their journey to become influential members of the fire service who create lasting, positive change in their communities.

Moving Forward

I hope this chapter has inspired you to become a stronger, more compassionate, and self-disciplined leader. I could go on for pages and pages about the examples of love, grace, and temperance in how I lead my fire department. However, the takeaway message is important: the more compassionate, forgiving, and self-moderated you are as a leader, the more likely you will gain the respect, trust, and compassion of those you lead.

No matter what type of leader you are, including love, grace, and temperance could change and completely transform the department you lead. As a leader, thinking about the greater good and your community should be at the top of your list.

I look forward to seeing even more fire department leaders, new fire chiefs, and firefighters exercising compassion, understanding, and self-discipline, even in the face of the harrowing events I have described. Being a part of the fire service has been one of the greatest joys of my life, but being able to lead and create a family atmosphere within the fire department has been all the more significant to my life, making even the most challenging professional moments worth experiencing.

If you embrace the lessons expressed in this final chapter, you will no doubt become the fire service leader you are destined to become. Embrace the power to change, elevate, and support those around you, through which you can change the lives the entire community, as well as those in your department.

As part of the fire service, there is no greater good than knowing that your service to the community has changed their lives for the better and impacted them in a way that often is beyond words. With the inclusion of love, grace, and temperance, you will undoubtedly have an effect on your community that will forever be remembered—and hopefully carried on in the generations to come, through those you lead.

Chapter 16 Review Questions

1. How does temperance aid in managing emotions and stress within the fire service?
   a. By encouraging impulsive reactions to emotional situations
   b. By suppressing all emotional responses
   c. By allowing an approach that balances compassion and stoicism
   d. By displaying erratic and unpredictable behavior

2. What role does trust play in fostering compassionate communication within a fire department?
   a. Trust is not important in this context.
   b. Trust builds solid relationships and encourages open communication.
   c. Trust hinders communication among firefighters.
   d. Trust causes conflicts and misunderstandings.

3. What will embracing the lessons in this chapter enable a fire service leader to do?
   a. Remain stagnant and unchanging
   b. Foster negativity and conflict within the community
   c. Become a leader devoid of compassion
   d. Change, elevate, and support those around them

4. What should be the ultimate goal of conflict resolution within a fire department?
   a. To create discord and division among team members
   b. To find ways to provoke more conflicts
   c. To ignore the needs and concerns of the parties involved
   d. To reach a mutually beneficial situation for both parties involved

5. How can a leader encourage more compassionate communication within the fire department?
   a. By setting strict rules and boundaries
   b. By fostering a team atmosphere based on empathy and compassion
   c. By ignoring the emotions and concerns of the firefighters
   d. By avoiding discussions and debriefing sessions

6. Encouraging personal development and growth in a fire department does not benefit the overall success of the organization.
   a. True
   b. False

7. Losses and emergencies within the fire service can test one's compassion, grace, and temperance.
   a. True
   b. False

8. It is essential for fire service leaders to know that their service has impacted the community in a positive way.
   a. True
   b. False

9. Mastering temperance can help when making decisions in high-stress environments.
   a. True
   b. False

10. Self-discipline as a leader impacts only your immediate team and not the community beyond the fire department.
    a. True
    b. False

11. How can leaders in the fire service integrate love and understanding to navigate challenging situations effectively?

12. How does cultivating forgiveness and understanding contribute to the mental health of the firefighters in a fire department?

13. What role does vulnerability play in enhancing compassionate communication within the fire service?

14. How can a leader demonstrate temperance when dealing with challenging situations in the fire service?

15. Why is it important for leaders in the fire service to show empathy and compassion toward their team members?

CONCLUSION: WILL YOU CARRY THE FIRE?

Many mistakenly define leadership as a function of an individual's charisma, a collection of behaviors and skills, or approaches and contingencies. These ideas are not supported by experience, as this book details. By contrast, leadership is a collaborative relationship between a leader and followers. It occurs in organizational cultures where leaders and followers are sufficiently united in values and trust each other to engage in joint activity.

Trust is truly more than a construct; it is, in fact, the foundation of the fire service. Without trust, the team suffers, and so will the public who we all serve as firefighters. As a leader in the industry, you must set the example. Be the leader with integrity who your followers can look up to. Become the fire service leader who you need to be to encourage your staff and demonstrate that your department has power and a role in society.

The chapters of this book have presented various leadership strategies that you can abide by—allowing you to curate a leadership style that works for you and your team, as well as meeting your goals for innovation. In addition, the importance of accountability and responsibility have been stressed, both within the team and individually, since you would struggle without a proper expectation for your staff—especially in the face of crises that require your leadership.

Above all else, the message shared in this book comes down to one critical decision you must make—not only for yourself and the firefighter you want to become but also for the up-and-coming firefighters you mentor and serve: Will you carry the fire and show your team the way?

As a leader—whatever your rank within your organization—you must decide to take on the task. Are you willing to demonstrate courage? Are you willing to live by your values? Are you committed enough to create a values-driven, mission-focused organization?

If your answer is yes, then you are ready to carry the fire. However, remember that this decision is just the start of your leadership journey. Every day on the job, you have an opportunity to show up and exhibit the teachings outlined in this book. You have the chance to make an impact, to evolve your team, and to inspire the next generation of firefighters.

To carry the fire is to hold oneself accountable, to follow your drive and passion for excellence, and to embrace resilience—even in the most troubling times. Your decision to carry the fire encourages others, igniting the spark for them to take the torch.

While not an easy path, I hope you do choose to carry the fire. By cultivating an environment of support and empowerment and working with your team, you can create the ideal department that contributes and succeeds together. Never forget that the flame burns within each of us—which is why we have all been called to the fire service over the years.

May your flame of leadership inspire others and may your leadership journey pave the way for a culture where your mission drives your team to leave behind a positive, lasting legacy for the fire service.

END-OF-CHAPTER REVIEW QUESTIONS ANSWER KEY

Chapter 1: Exploring Today's Leadership Methodologies

1. What are the two main categories of leadership?
   C. Transactional and transformational leadership

2. What type of leader sets expectations and parameters for punishments or rewards?
   C. Transactional leader

3. Which of the following is a benefit of successfully applying meta-leadership?
   C. Building sustainable collaborations

4. Why is ethical leadership important for organizations in today's world?
   D. To emphasize corporate social responsibility

5. Which of the following are the three important components of the meta-leadership framework?
   B. The person, the situation, and the connectivity

6. Passive management by exception in transactional leadership involves addressing issues immediately.
   B. False

7. Sinek's thought leadership emphasizes starting with why to become a purpose-driven leader with authentic communication and emotional engagement.
   A. True

8. It is not important to know how to apply leadership theories and philosophies.

   B. False

9. Under purpose-driven leadership, the *why* is the central driving force behind leadership success.

   A. True

10. The text focuses on transformational leadership theories, rather than traditional leadership theories.

    B. False

11. What qualities make an authentic leader stand out in an organization?

    Authentic leaders excel in self-awareness, transparency, relational empathy, consistency, and resilience, fostering trust, engagement, and a positive work environment.

12. How can servant leadership empower firefighters and staff to seek improvement in their skill set?

    By promoting a culture of service and continuous improvement, servant leadership encourages individuals to go beyond their expectations and reach for excellence in all their endeavors.

13. Why do ethical leaders play a crucial role in setting organizational standards?

    Ethical leaders serve as role models for ethical behavior, establish clear values, and ensure that the organization follows a moral code, thereby setting a standard for ethical conduct within the organization.

14. Why is it important for a servant leader to have empathy and compassion for others?

    Having empathy and compassion allows them to understand their team members' perspectives and needs, fostering a supportive work environment.

15. What are some potential benefits of successfully applying the meta-leadership framework?

    Benefits include enhanced team collaboration, improved decision-making skills, crisis response preparedness, and the ability to provide sustainable solutions.

Chapter 2: Organizational Culture

1. Which of the following is NOT a way to prevent resistance to change in the organizational culture?

   D. Be a mentor or implement a reverse mentorship program

2. How does mission-driven culture make the ability for feedback and employee development more efficient?

   D. By appropriately measuring operational outcomes and recognizing performance

3. Which of the following describes how to keep team members in the loop during the change process?

   D. Mentoring them, discussing apprehensions, and maintaining communication

4. How does mission-driven culture help leaders focus on their priorities within the organization?

   C. By ensuring strategic alignment and balancing priorities

5. How does constructivism differ from behaviorism in promoting change in organizations?

   C. Constructivism encourages learners to be active participants in the learning process.

6. Cultural change can be implemented effectively through constructivism by ensuring that people understand the values of the organization.

   A. True

7. Mission-driven cultures in fire departments can lead to an increase in job satisfaction for firefighters.

   A. True

8. A leader who believes in the autocratic approach may face resistance from subordinates during change.

   A. True

9. Constructivism encourages learners to explore and take part in an active learning process.

   A. True

10. A key aspect of mission-driven culture is that it is driven by individual goals and the overall mission of the organization.

    A. True

11. How important is the role of leaders in shaping the culture and principles of a fire department?

    The role of leaders is crucial as they define the organizational culture through decision-making and management approaches. By shaping the culture, they influence how the public perceives the fire department.

12. How does the fire and emergency services culture promote continuous learning and growth among its members?

    The fire and emergency services culture fosters continuous learning and growth by encouraging ongoing training, skill development, and knowledge sharing among its members. By valuing professional development, mentorship, and the exchange of expertise, the culture empowers firefighters to improve their skills, adapt to new challenges, and enhance their effectiveness in serving the public.

13. What are the key factors to consider when formulating a strategy for implementing change within an organization?

    For a change implementation strategy, it is essential to consider cultural norms, communication methods, team dynamics, and potential resistance to ensure a successful and sustainable transition.

14. How does the fire and emergency services culture inspire new recruits to embody the values of teamwork and service to others?

    Through traditions, role modeling by senior members, and a strong sense of community responsibility, recruits are inspired to embrace the core values of the organization and contribute positively to the team dynamic.

15. How does mission-driven culture impact the organizational structure of fire departments?

    Mission-driven culture decentralizes decision-making, allowing leaders to guide intentions and provide expectations while giving staff members the freedom to make intelligent decisions. This leads to a more flexible and adaptive organizational structure.

Chapter 3: Dilemmas, Dissonance, Ethics, and Legal Concerns in the Fire Service

1. Ethics codes become incongruent and incompatible across different states or regions because of which of the following?

   C. Varying legal requirements

2. Which of the following is a critical legal challenge that can happen following the service performed by fire stations in the community?

   E. Negligence, property damage, and wrongful death lawsuits

3. Why must ethics become an organizational priority in the 21st century?

   D. To succeed in a competitive environment

4. Which of the following is a significant concern in volunteer and combination fire departments regarding staffing?

   D. Low retention rates

5. What kinds of factors influence the likelihood of a fire?

   D. All of the above

6. Legal concerns can be minimized with oversight, accountability, and education in the fire service.

   A. True

7. Heath suggests that applying upstream thinking in organizations can help prevent legal issues in the fire service.

   A. True

8. Property damage, negligence, and wrongful death lawsuits are not common legal issues in the fire service.

   B. False

9. Preparing a fire department with proper leadership can prevent most legal concerns.

   A. True

10. Communication of expectations and reasons by leadership is not important in fire suppression efforts.

    B. False

11. What challenges do fire service agencies face in maintaining adequate staffing levels, especially in smaller towns?

    Budget cuts and limited resources have led to understaffing, forcing some fire departments to downsize or close completely; these also impact training, equipment maintenance, and ultimately endanger both firefighters and communities.

12. How do funding sources and levels for fire departments vary, and what challenges do they face in managing fire threats?

    Funding sources and levels for fire departments vary by country. Fire departments receiving constrained funding often face challenges in managing known fire threats owing to resources that are inadequate to meet all department objectives.

13. In what ways can oversight and accountability help in addressing legal issues within fire departments?

    Oversight and accountability ensure that proper protocols are followed, minimizing the risk of legal issues arising.

14. Why should ethics be considered a priority rather than just a legal or moral obligation for organizations?

    Ethics should be considered a priority for organizations because it strengthens the organizational culture, aligns with values such as accountability and professionalism, and contributes to the overall success and reputation of the organization in a competitive environment.

15. What legal issues might firefighters face in the field?

    Firefighters may face property damage, negligence, and wrongful death lawsuits stemming from the service they provide in the community.

Chapter 4: Managing the Brand

1. Why is it important for the fire service to be involved in the community?

   D. To help civilians understand the critical role of the fire service

2. How can a budget contribute to optimizing brand recognition efforts?
   B. By investing in digital marketing tactics

3. What is community risk reduction?
   C. The culture of the fire service

4. How can firefighters promote their brand?
   C. By raising fire safety awareness and preventing accidents

5. How can the fire service enhance its brand awareness?
   C. By having a logo with a strong message or symbolic value

6. The main goal of brand management is to maintain a positive perception among the target audience.
   A. True

7. Marketing strategies for the fire service should focus primarily on the tangible services provided rather than the emotions associated with them.
   B. False

8. Customers are more likely to stay loyal to a brand they trust and recognize.
   A. True

9. Brand recognition can be enhanced through marketing and advertising activities.
   A. True

10. Supervisory duties are not part of the responsibilities of management in a fire station.
    B. False

11. How does brand consistency in messaging and communication help a fire department to establish credibility and trust with its stakeholders?
    Consistent messaging across various platforms and interactions demonstrates reliability, transparency, and authenticity, reinforcing the organization's commitment to meeting stakeholders' needs and expectations.

12. What role does community involvement play in public health practices and firefighting strategies?
    Community involvement plays a vital role in public health practices and firefighting strategies by actively engaging individuals in identifying

issues, designing solutions, and sharing outcomes within the community. This collaboration leads to more effective problem-solving and enhances the overall impact of public health and firefighting efforts.

13. What is brand equity, and how does it impact a fire department's pricing and customer perception?

    Brand equity refers to the value a brand provides to goods or services. This value influences customers' willingness to pay a premium for products associated with trust and recognition, ultimately enhancing the overall brand image.

14. In what ways can firefighters use social media to connect with their community and demonstrate brand value?

    Firefighters can use social media to provide real-time updates during emergencies, inform the public about humanitarian initiatives, and facilitate coordination among multiple parties involved in disaster response. Engagement with customers allows them to see the brand value.

15. Why is maintaining a well-structured brand voice important for a fire department across various media platforms?

    Consistent brand voice reflects organizational culture and values, helping to establish a recognizable identity that resonates with customers and stakeholders across different communication channels.

Chapter 5: The High Cost of Low-Fidelity Leadership

1. How can well-defined core values benefit an organization?

   C. They assist in defining criteria for actions and attitudes needed to attain goals

2. Which of the following is critical for building a solid and successful organizational culture?

   C. Integration of a compelling narrative

3. How does the chapter suggest influential leaders can transform organizations?

   C. By reinforcing organizational values from the inside out

4. How can a clear vision and mission statement help build an organizational structure?

   E. It delays the achievement of organizational goals.

5. Why is it important to have a reward system in place?

   C. To encourage desirable behaviors and actions

6. Integrity in leadership doesn't have any impact on organizational culture or employee behavior.

   B. False

7. Vision and mission statements in organizations play a minimal role in guiding day-to-day decisions and interactions.

   B. False

8. Bridging communication gaps is not essential for project success in a fire department.

   B. False

9. Actions should align with the mission and vision statements to solidify organizational culture.

   A. True

10. In the example, the new fire chief demonstrated integrity and value, which ultimately led to increased motivation and commitment among the firefighters.

    A. True

11. How does low-fidelity leadership differ from high-fidelity leadership?

    Low-fidelity leadership focuses on mistakes and failures, while high-fidelity leadership emphasizes doing things correctly and learning from success stories. High-fidelity leaders take a positive and proactive approach to leadership.

12. In what ways can leaders demonstrate integrity in their actions to build trust with employees?

    Leaders can demonstrate integrity by keeping promises, making fair and ethical decisions, recognizing employees' achievements, and fostering a collaborative work environment. By displaying honesty and reliability, leaders can earn the trust of team members.

13. Why is it crucial for organizations to recruit individuals who align with their culture?

    Hiring individuals who share or can adopt the organization's values builds a cohesive culture. Studies show that culturally fit candidates are more likely to stay longer and contribute positively to the organization.

14. Why is it challenging to restore trust in the workplace once it has been broken?

    Restoring trust in the workplace is difficult because once broken, there will be skepticism and uncertainty among employees. It can take a significant effort and time to rebuild trust after it has been damaged.

15. How does a positive organizational culture contribute to the success of a fire department?

    A positive organizational culture fosters teamwork, motivates firefighters, and encourages commitment, ultimately leading to improved performance and effectiveness during critical situations.

Chapter 6: Discipline

1. Why is it important for a leader to accept blame and strive toward corrective measures, even if they are not personally at fault?

   E. To maintain workplace balance

2. What is the impact of lack of discipline within an organization?

   C. Results in chaotic behavior

3. What is the importance of learning from negative situations and tackling obstacles head-on?

   B. It allows for continuous improvement.

4. How does providing value to others contribute to personal development?

   C. It fosters a positive impact on others and oneself.

5. How do disciplined employees differ from nondisciplined employees in terms of organizational regard?

   C. Disciplined employees are regarded more positively by their colleagues.

6. Accountability is about recognizing that you are in control of your life and your actions.

   A. True

7. Integrity is not considered an essential quality for a successful leader.

   B. False

8. Subordinates are likely to trust and approach their managers more if the managers exhibit dishonest behavior.

   B. False

9. A good leader stands out by taking total accountability for their choices and the consequences.

   A. True

10. Leadership exemplifying discipline and accountability may not impact employees' behavior.

    B. False

11. What key qualities do leaders need to demonstrate to create a successful and ethical work environment?

    Leaders need to demonstrate care, integrity, trustworthiness, effective management skills, respect for team members, adherence to organizational values, and a commitment to fostering a positive and ethical work environment.

12. What are the consequences of a lack of discipline within an organization?

    The consequences of a lack of discipline in an organization include sabotaging core performance potential, destroying corporate integrity, deviating from the mission and values, and undermining leadership's ability to reflect the organization's ideals fairly.

13. How can good leadership favorably influence employee behavior and promote a positive work environment?

    Good leadership fosters trust, respect, and accountability among workers, leading to a positive work environment where employees feel valued and motivated to work toward common goals.

14. What does it mean to be the president and CEO of your life in terms of accountability?

    Being the president and CEO of your life means taking control, accepting responsibility for choices, and realizing that you are in charge of the outcomes of your actions and relationships.

15. How does providing employees with opportunities to atone for their mistakes contribute to a positive organizational culture and climate?

    Allowing employees to atone for their mistakes contributes to a positive organizational culture by enabling them to learn, improve, and feel valued members of the team, under the guidance of accountable and disciplined leadership.

Chapter 7: Accountability—the Antidote to Chaos

1. Which of the following is the responsibility of an organizational leader?

   C. To build relationships and create trust

2. How should a leader build trust and fidelity with an organization's membership?

   C. Consistently demonstrating accountability

3. Which of the following is a necessary step to effect change in the accountability process?

   D. Documenting both successful and failed attempts

4. Which of the following is the primary focus of the Accountability "P" continuum?

   C. Taking ownership of outcomes

5. What is organizational freelancing?

   C. Irresponsible actions within a permissive organizational environment

6. Authentic leadership involves avoiding learning from failures.

   B. False

7. Fear of failure should be an impediment to action.

   B. False

8. Leaders should avoid forming cross-functional teams to utilize all employees' ideas and talents.

   B. False

9. Constructive feedback should be vague and open to interpretation to encourage employee self-reflection.

   B. False

10. Leaders who invest in holding employees accountable for their actions and providing feedback show that their actions and behaviors impact the organization's mission.

    A. True

11. How can forming cross-functional teams benefit an organization in terms of creativity and problem-solving?

    Forming cross-functional teams can benefit an organization by fully utilizing employees' ideas and talents, improving communication, capitalizing on opportunities, and developing practical solutions to complex situations.

12. How does the aphorism "Don't practice until you get it right; practice until you can't get it wrong" resonate with the mindset of firefighters regarding failure?

    The aphorism underscores the resilient mindset required in firefighting, where failure is seen as a practice opportunity to perfect skills, learn, and evolve into greater professionals in the field.

13. How can embracing failure in a controlled environment contribute to developing adaptive solutions in the fire service?

    Embracing failure in a controlled environment allows for experimentation to see what works and what doesn't, leading to the development of adaptive solutions to nonconforming problems.

14. How can accepting and valuing failure foster a more resilient organizational culture in the fire service?

    Accepting and valuing failure can promote a culture of learning, growth, and adaptation, enhancing organizational resilience and readiness to face challenges.

15. How can organizations prevent normalization of deviation and promote a culture of accountability?

    Organizations can prevent the normalization of deviation and promote a culture of accountability by setting clear expectations, providing regular feedback, holding individuals accountable for their actions, and fostering a culture of continuous improvement and learning. Leadership commitment and consistent reinforcement of accountability are also essential.

Chapter 8: The Importance of Firefighter Health

1. Why is mental health described as the foundation of resilience for firefighters?

   C. To withstand high-stress environments

2. Why is crew resource management (CRM) considered a foundational aspect of psychological safety but not the end goal?

   C. Psychological safety encompasses broader measures beyond CRM.

3. Why is education considered an important strategy for preventing moral injuries in the fire service?

   C. It assists team members in understanding impacts and addressing challenges.

4. What approach does Hudson advocate to destigmatize mental health issues among firefighters?

   C. Normalize discussing depression, anxiety, and other issues.

5. Which of the following is critical when dealing with moral injuries in the fire service?

   C. Ensuring well-being and safe stress management

6. Creating a supportive environment within the fire service allows individuals to speak up without fear of retaliation.

   A. True

7. Access to counseling services, support groups, and mental health training is not important in ensuring a psychologically safe work environment.

   B. False

8. Peer support groups can be beneficial in promoting a more robust team culture and helping individuals feel heard and valued.

   A. True

9. Strong leadership is not crucial in maintaining a psychologically safe work environment.

   B. False

10. Implementing evidence based practices such as counseling and cognitive behavioral therapy can help manage moral injuries in the fire service.

    A. True

11. How does effective leadership foster a mentally healthy work environment for firefighters?

    Effective leadership sets the tone for a psychologically healthy workplace by promoting active listening, clear communication, and constructive feedback and creating a sense of belonging and trust among team members to support mental well-being.

12. Why is it crucial to suppress stigmas around mental health in a fire service team to foster psychological safety?

    Suppression of stigmas allows team members to feel safe discussing their mental health without fear of judgment. This enables them to seek help and support when they might not otherwise.

13. In what ways can understanding emotional impact benefit firefighters in managing stress and anxiety?

    Recognizing emotional effects helps firefighters to address stress, anxiety, and PTSD, allowing them to take necessary steps to manage their mental well-being effectively.

14. In what ways can continuous evaluation and improvement benefit the overall effectiveness of a fire department?

    Through continuous evaluation and improvement, leadership can identify and address shifts in organizational culture, ensure ongoing support for team members, and maintain a strong, cohesive, and effective team.

15. How can moral injuries impact a firefighter's mental health and overall well-being?

    Moral injuries can lead to poor mental health, anxiety, depression, PTSD, and decreased job satisfaction among firefighters. It can also affect their relationships with others and lead to a sense of isolation or detachment.

Chapter 9: Five Laws of Leadership

1. How should a visionary leader modify their vision as the organization evolves?

   B. By sharing a brand-new future for the team to work toward

2. Which of the following best describes ownership?

   C. Taking responsibility and pride in actions

3. Which of the following is important for successful fire service organizations?

   C. Taking responsibility for actions and consequences

4. What is the relationship between honesty, trust, and leadership?

   C. Honesty builds trust and legitimacy.

5. What should organizations use to determine an appropriate course of action in the face of noncompliance or deviation?

   C. Judgment, discernment, and organizational values

6. Leaders need to inspire and set an example for their team by demonstrating the right way.

   A. True

7. Continuously training and gaining experience will not significantly impact an individual's competence and compliance in their profession.

   B. False

8. Leaders should not be held accountable when an endeavor fails, as it is a team effort.

   B. False

9. Organizations should trust and support leaders' intent, even if noncompliance is a conscious decision with no justifiable reason.

   B. False

10. Blaming other team members can foster a nurturing culture in the workplace.

    B. False

11. How is honesty related to the concepts of positivity and integrity in leadership?

    Honesty promotes a positive work environment by encouraging open communication and upholding ethical standards, which are crucial for maintaining integrity.

12. How can a leader inspire loyalty and commitment from team members?

    By leading by example, demonstrating proficiency, and providing support and guidance, leaders can inspire loyalty and commitment from their team.

13. What is the significance of compliance with organizational rules and regulations supported by mission and values in promoting a healthy work environment?

    Compliance with organizational rules ensures a universal standard is observed by all members. When rules are aligned with the organization's mission and values, they become more than just words on paper (or a screen) and contribute to a positive work culture where everyone works toward a common goal.

14. How can a leader balance accountability for both successes and failures within the team?

    Leaders must take responsibility for outcomes, provide guidance and support, and encourage the workforce to follow agreed steps while monitoring progress.

15. In what ways can a leader promote a culture of accountability and ownership within the team, especially when handling mistakes?

    Leaders should take responsibility for mistakes, encourage team members to own up to their errors, and foster an environment where discussions about mistakes lead to learning and growth.

Chapter 10: Developing the Leadership Ethos of Your Team

1. How can mentorship programs contribute to the future of the fire service?

   D. By providing knowledge and experience to mentees

2. Which of the following best describes the joy of leadership when it comes to motivating employees?

   B. Evolving and adapting to meet the team's needs and desires

3. How can the impact of a mentorship program be optimized through support?

   B. By offering adequate support to both the mentor and mentee

4. Why is it crucial for mentors to be ethical and honorable individuals in a mentorship program?

   C. To ensure integration into the designed program

5. Why is it essential for senior staff in the fire service industry to design mentorship programs that meet specific objectives?

   D. To align the mentorship program with the organization's goals

6. Mentorship in the fire service is not considered essential for preparing firefighters for their future career.

   B. False

7. An experienced leader in the fire service can provide knowledge, expertise, and on-the-job training to a new firefighter.

   A. True

8. The stress and emotional turmoil of the firefighter role can significantly impact an individual's life and mental stability.

   A. True

9. Building a relationship with an active firefighter in the firehouse can create a culture of exclusion among firefighters.

   B. False

10. A mentoring program should set achievable goals and objectives as the first step.

    A. True

11. What role does communication play in the success of a mentorship program?

    Communication is vital to maintain contact between the mentor and mentee, enabling them to discuss experiences and ensure the program's success.

12. How can effective leadership evolve to meet the diverse needs and desires of a team?

    Effective leadership evolves by constantly adapting to meet the varying needs and desires of the team members, allowing for personalized motivational approaches that align with individual aspirations and organizational objectives.

13. Why is leadership not solely about leading an entire team but also about developing individuals into leaders themselves?

    Leadership requires investing in individuals to help them become leaders, by guiding them in the right direction and fostering their growth and potential for personal and professional development.

14. Why is the carrot-and-stick approach not successful in motivating fire service personnel?

    The carrot-and-stick approach is not successful in motivating the fire service community because traditional leadership methods that involve punitive actions or rewards for desired behaviors do not address the intrinsic motivation of firefighters, who are driven by values such as altruism and humanitarianism.

15. How can mentorship programs impact talent management in fire service organizations?

    Mentorship programs can attract and retain a diverse talent pool by embracing diversity and inclusivity and appreciating individual contributions within the organization's regulations.

Chapter 11: Developing Community and Organizational Equity

1. Which of the following is a challenge regarding communication preferences in a multigenerational workforce?

   D. Understanding differing preferences is necessary to find common ground.

2. Which of the following should be the foundation for every fire service leader and member to create a diverse and inclusive organizational culture?

   D. Incremental progress

3. How can a healthy mindset in an organization contribute to an individual's mental and physical health?

   C. By emphasizing the importance of mental health alongside physical health

4. According to *Managing Workplace Diversity and Inclusion*, what is the importance of inclusivity?

   D. It fosters a supportive environment for team members.

5. Which of the following best describes the process of creating a diverse, inclusive organization in the fire service?

   B. A marathon

6. Diversity is concerned with data, whereas inclusivity is about the ecosystem within organizations.

   A. True

7. Diversity refers to variation in demographic characteristics such as race, ethnicity, and sex in the workplace.

   A. True

8. Millennials prefer face-to-face interaction over digital technologies and social media.

   B. False

9. Creating a diverse and inclusive fire organization should primarily rely on adhering to federal law.

   B. False

10. A good leader should always have a clear vision and plan for the future.

    A. True

11. Why is it important to address negative stereotypes in a multigenerational workforce?

    Addressing negative stereotypes fosters a supportive environment, builds confidence among team members, and promotes harmonious staff interactions.

12. How do stress levels impact an individual's physical health and well-being?

    Higher stress levels are associated with a greater risk of cardiovascular disease, chronic pain, inflammation, and other health issues. Thus, managing stress is important for overall well-being.

13. How can a fire department make its recruitment process more inclusive of underrepresented groups?

    A fire department can bolster inclusivity by collaborating with community organizations and schools, using inclusive language and images in job advertisements, and actively participating in events such as job fairs and career days.

14. Why is consistency more important than intensity in addressing diversity and inclusivity in the fire service?

    Consistency in addressing diversity and inclusivity ensures long-term progress in becoming more representative of the communities served because small, deliberate steps are more impactful than intense but sporadic efforts.

15. How can a leader effectively manage differing cultural expectations among different generations?

    A leader can effectively manage differing cultural expectations by promoting respect, understanding, and open communication to bridge gaps and enhance collaboration within the team.

Chapter 12: Listening and Crucial Conversations

1. What is the default mode of communication for Generation Z?

   C. Using messaging platforms and hosting digital meetings

2. Which form of communication do older generations often prefer?

   D. Speaking face-to-face

3. Why is actively listening important for strong communication?

   C. It aids in decision-making and understanding the situation.

4. Which of the following statements best describes communication?

   B. It involves giving, receiving, and sharing information.

5. How can communication skills help leaders to achieve milestones efficiently?

   D. By establishing a sound argument

6. Putting aside gadgets and actively participating in conversations can aid in effective communication, in both firefighting and project planning.

   A. True

7. Encouraging a team atmosphere in a fire service organization is not necessary for improving emergency response skills.

   B. False

8. Using online platforms such as Zoom or Teams could be ideal for people to talk, share, and learn about their tasks, supporting one another during crises in a collaborative environment.

   A. True

9. Active listening is an essential component of strong communication as it helps in understanding situations and making effective decisions.

   A. True

10. Strong communication in any organization primarily relies on three main attributes: comprehension, compassion, and empathy.

    A. True

11. What are some key considerations for organizations looking to create a communication strategy that effectively addresses the needs of diverse generations in the workplace?

    Key considerations include recognizing and respecting generational differences in communication preferences, incorporating a mix of traditional and digital communication channels, and promoting a culture of understanding and flexibility in communication practices.

12. What are some consequences of operating within a siloed organization that lacks cross-collaboration?

    In a siloed organization, the lack of cross-collaboration can lead to a fragmented work environment, miscommunication, and a sense of isolation among team members. This may result in decreased efficiency, ineffective problem-solving, and a negative impact on morale.

13. In what ways can setting goals for the organization and fire department leaders promote a team atmosphere in a fire service organization?

    Setting clear goals for the organization and fire department leaders can encourage teamwork by providing a common purpose for all members to work toward, thereby promoting unity and collaboration within the department.

14. Why is it important for a fire service organization to prioritize team cohesion and communication even during challenging times when remote work is necessary?

    Prioritizing team cohesion and communication within a fire service organization during challenging times and when working remotely is essential to maintain unity, coordination, and effectiveness in operations, ensuring that members do not operate in isolated silos but instead work together toward common goals.

15. How can understanding different generations' preferred methods of communication improve overall communication within a team?

    By recognizing and adapting to varying preferences such as face-to-face communication for older generations, email/text for millennials, and messaging platforms/digital meetings for Generation Z, a team can bridge communication gaps effectively.

Chapter 13: Hiring, Interviewing, and Onboarding

1. What is the primary goal of training and development programs for staff members?

   D. To train employees to perform in a way that is expected of them

2. What is essential for the long-term retention of staff members, both new and old?

   D. Providing both immediate and continual training

3. Why is it important to outline the organizational culture during the onboarding process?

   C. To help new staff members integrate easily with existing staff

4. What is the primary goal of the multifaceted approach to recruitment in fire departments?

   E. Attracting skilled professionals via various channels

5. Which of the following best describes the ideal focus of recruitment efforts?

   D. Finding individuals passionate about defending their community

6. Self-control is not necessary for becoming a command-resonant leader.

   B. False

7. Employees who are optimistic about their future in the fire service or EMS are more likely to attract others to join the profession.

   A. True

8. Self-awareness is not a crucial trait for developing command resonance.

   B. False

9. It is important for organizations to align their recruitment strategies with their long-term goals and objectives.

   A. True

10. Admitting mistakes and being accountable for decisions helps a leader gain trust and authenticity.

    A. True

11. How does practicing forgiveness benefit team dynamics in a leadership role?

    Forgiveness promotes a culture of understanding, empathy, and growth within a team, fostering stronger relationships and a positive work environment.

12. What key factors do recruiters consider when hiring for a tough job such as firefighting?

    Recruiters look for physical fitness, mental resilience, problem-solving skills, teamwork abilities, and a passion for the firefighting profession.

13. What are some examples of locations where fire service professionals could set up recruitment booths?

    Fire service professionals could set up recruitment booths at local farmer's markets, music festivals, houses of worship, military outplacement job fairs, or community events to attract potential applicants.

14. What can fire service recruiters do to target messaging at potential recruits in the community?

    Recruiters can target messaging at the family, extended family, and friends of current personnel and retirees to attract individuals who already have a foundational understanding of the career and may be interested in following in the footsteps of those individuals.

15. What role does authenticity play in command-resonant leaders?

    Authenticity is essential for command-resonant leaders to be genuine, live by their values, showcase courage, and practice forgiveness.

Chapter 14: Stoicism, Resilience, and the Fire Service

1. Why is it important for individuals to exercise both the analytical and logical sides of the brain when making decisions?

   C. To cultivate the skills of critical thinking and decisive action

2. What does the mental framework of stoicism encourage firefighters to see?

   B. Circumstances of opportunity

3. How does engaging in critical thinking regularly benefit leaders?

   D. Encourages contemplation of arguments and available information

4. Why is questioning one's assumptions important in critical thinking?

   D. To reconsider positions based on new information

5. Which of the following are two essential characteristics of top firefighters and fire officers?

   B. Critical thinking and decisive action

6. Responsive behavior in decision-making involves considering both short- and long-term outcomes.

   A. True

7. Resilience can be developed over time rather than being an innate trait.

   A. True

8. Critical thinking is about being emotional and impulsive in making decisions.

   B. False

9. Critical thinking and decisive action should be fueled by asking important questions.

   A. True

10. Engaging in critical thinking regularly allows leaders to reconsider their positions based on evolving information.

    A. True

11. How does stoicism help firefighters develop situational awareness?

    Stoicism helps firefighters develop situational awareness by improving the desire to recognize challenges and obstacles with a cue-primed response model based on confidence, not fear.

12. What does Seneca's statement "Excellence withers without an adversary" mean in the context of stoicism and firefighting?

    Seneca implies that facing challenges is essential for growth and excellence. In the fire service, overcoming adversities through stoic virtues such as courage and wisdom leads to personal and professional development.

13. How does having resilience and stoicism benefit firefighters in their profession?

    Having resilience and stoicism benefits firefighters by allowing them to maintain an indomitable spirit in irrational or unfortunate circumstances, avoiding overreactions, and developing discernment and balance in an ever-evolving profession.

14. What is the difference between being responsive and reactive (reactionary) in decision-making?

    Being responsive involves calculated, decisive actions by anticipating future outcomes, while being reactive involves making decisions based on past events.

15. Why is it important for officers in the fire service to learn the stoic principle of wisdom?

    It is important for officers in the fire service to learn the stoic principle of wisdom because it allows them to feel comfortable realizing they do not have to have all the answers but have a desire to learn and integrate from other examples.

Chapter 15: The Importance of Strategy and Trust

1. Which of the following best describes self-serving motives?

   D. Normal and healthy but need to be recognized and understood

2. In the fire service, why is trust considered the building block for critical thinking and strategic planning?

   C. Trust enables critical thinking and strategic planning to be implemented effectively.

3. How should strategic thinking and strategic planning be prioritized in a fire service organization?

   C. Strategic thinking is more crucial in today's rapidly changing environment.

4. Why is information sharing crucial for organizational strategic development in the fire service?

   E. Trust developed from information sharing can confirm the agency's goals.

5. Which of the following is a best practice for overcoming the challenges of self-interest versus team orientation?

   C. Recognizing and understanding self-serving motives

6. Strategic planning alone is sufficient for organizations to address environmental changes effectively.

   B. False

7. The presence of trust is crucial for ensuring that critical thinking and strategic planning are optimally implemented in any organization, particularly in the fire service.

   A. True

8. Leadership and trust are not crucial elements for the success of a fire service culture.

   B. False

9. Communication trust involves the willingness to share information, tell the truth, admit mistakes, and keep secrets.

   A. True

10. Strategic planning and thinking in a fire service agency are completely independent of each other.

    B. False

11. How can understanding the concept of trust dynamics help fire service professionals foster a culture of trust within their organizations?

    By comprehending the different dynamics and types of trust, fire service professionals can implement strategies to cultivate and strengthen trust within their organizations. This knowledge allows for better communication, clearer expectations, and collaborative decision-making, ultimately enhancing team cohesion and performance.

12. How do traditionalism and cynicism act with regard to instilling trust within a team?

    Traditionalism and cynicism can act as barriers that hinder trust-building efforts because they lead to resistance to change and a lack of belief in leadership.

13. Why is trust considered a building block for effective strategic planning and strategic thinking?

    Trust fosters open communication, collaboration, and a sense of security within a team, which are all essential elements for successful strategic planning and thinking.

14. How can fire service leaders integrate strategic thinking into their decision-making processes effectively?

    Fire service leaders can integrate strategic thinking by encouraging upstream thinking, envisioning future scenarios, and aligning departmental visions with organizational goals. By focusing on core values, mission statements, and innovative problem-solving, leaders can design solutions that anticipate and address challenges proactively.

15. How can fire service organizations balance the need for strategic planning and strategic thinking to adapt to evolving environments?

    Fire service organizations can balance the need for strategic planning and strategic thinking by integrating structured processes with innovative decision-making. By combining the foresight of strategic thinking with the actionable plans of strategic planning, organizations can adapt to changing environments while maintaining operational efficiency and competitiveness.

Chapter 16: Love, Grace, and Temperance in the Fire Service

1. How does temperance aid in managing emotions and stress within the fire service?

   C. By allowing an approach that balances compassion and stoicism

2. What role does trust play in fostering compassionate communication within a fire department?

   B. Trust builds solid relationships and encourages open communication.

3. What will embracing the lessons in this chapter enable a fire service leader to do?

   D. Change, elevate, and support those around them

4. What should be the ultimate goal of conflict resolution within a fire department?

   D. To reach a mutually beneficial situation for both parties involved

5. How can a leader encourage more compassionate communication within the fire department?

   B. By fostering a team atmosphere based on empathy and compassion

6. Encouraging personal development and growth in a fire department does not benefit the overall success of the organization.

   B. False

7. Losses and emergencies within the fire service can test one's compassion, grace, and temperance.

   A. True

8. It is essential for fire service leaders to know that their service has impacted the community in a positive way.

   A. True

9. Mastering temperance can help when making decisions in high-stress environments.

   A. True

10. Self-discipline as a leader impacts only your immediate team and not the community beyond the fire department.

    B. False

11. How can leaders in the fire service integrate love and understanding to navigate challenging situations effectively?

    Leaders in the fire service can integrate love and understanding by considering the feelings of all parties involved, understanding perspectives, and working toward solutions that address the needs and desires of everyone affected.

12. How does cultivating forgiveness and understanding contribute to the mental health of the firefighters in a fire department?

    Cultivating forgiveness and understanding creates a supportive environment where firefighters can forgive themselves for mistakes, leading to better mental health and overall well-being.

13. What role does vulnerability play in enhancing compassionate communication within the fire service?

    Vulnerability plays a significant role in enhancing compassionate communication within the fire service by allowing leaders to connect on a deeper level with their team, show authenticity in their emotions, and create an environment where open dialogue and trust are encouraged.

14. How can a leader demonstrate temperance when dealing with challenging situations in the fire service?

    Leaders can practice temperance by exercising moderation and self-discipline in their actions and decisions. This approach helps maintain composure, strategic thinking, and overall success in leading the fire department.

15. Why is it important for leaders in the fire service to show empathy and compassion toward their team members?

    It is important for leaders in the fire service to show empathy and compassion toward their team members to build trust, foster relationships, and ensure that firefighters feel valued, supported, and heard in challenging situations.

NOTES

1. Exploring Today's Leadership Methodologies

1. Peter G. Northouse, *Leadership: Theory and Practice*, 6th ed. (Thousand Oaks, CA: Sage Publications, 2013), p. 186.
2. Ibid., p. 195.
3. Ibid.
4. Ibid., p. 196.
5. Ibid.
6. Ibid., p. 421.
7. Ibid.
8. Ibid.
9. Ibid., p. 329.
10. Ibid., p. 185.
11. Ibid., p. 186.
12. Ibid., p. 192.
13. John C. Maxwell and Rob Hoskins, *Change Your World: How Anyone, Anywhere Can Make a Difference* (Nashville: HarperCollins Leadership, 2021), p. 61.
14. Dorothy L. Sayers, *Why Work? An Address Delivered at Eastbourne, April 23rd, 1942* (London: Methuen, 1942).
15. Ibid.
16. Northouse, *Leadership*, p. 219.
17. John F. Kennedy, "President John F. Kennedy's Inaugural Address," National Archives, January 20, 1961, https://www.archives.gov/milestone-documents/president-john-f-kennedys-inaugural-address.
18. Sayers, *Why Work?*

19. Northouse, *Leadership*, pp. 219–20.
20. Ibid., pp. 228–29.
21. Ibid., pp. 230–31.
22. Sayers, *Why Work?*
23. Ken Blanchard and Renee Broadwell, *Servant Leadership in Action: How You Can Achieve Great Relationships and Results* (Oakland, CA: Berrett-Koehler Publishers, 2018), p. 58.
24. Northouse, *Leadership*, p. 253.
25. Ibid., p. 254.
26. Ibid., pp. 266–68.
27. Ibid., p. 269.
28. Ibid., p. 99.
29. Ibid., p. 100.
30. Ibid.
31. Ibid.
32. Ibid.
33. Ibid., pp. 105–6.
34. Ben Ramalingam et al., "5 Principles to Guide Adaptive Leadership," *Harvard Business Review*, September 11, 2020, https://hbr.org/2020/09/5-principles-to-guide-adaptive-leadership.
35. Ibid.
36. Ibid.
37. Ibid.
38. Ibid.
39. Northouse, *Leadership*, p. 429.
40. Ibid.
41. Ibid., p. 433.
42. Ibid., p. 430.
43. Ibid., pp. 436–37.
44. Ibid., p. 437.
45. Leonard J. Marcus et al., *You're It: Crisis, Change, and How to Lead When It Matters Most* (New York: Public Affairs, 2019).
46. Ibid.
47. Ibid.
48. Ibid.
49. Ibid.
50. Simon Sinek, *Start with Why: How Great Leaders Inspire Everyone to Take Action* (New York: Penguin, 2009), p. 168.
51. Ibid., p. 39.
52. Ibid., pp. 39–40.

2. Organizational Culture

1. Keith Collins, "Firefighter Salary vs. Firefighter Value," FireRescue1, September 14, 2018, https://www.firerescue1.com/amu/articles/firefighter-salary vs firefighter-value-BXw51Tmp3LoKXtdA/.
2. Kris Blume, "Creating a Culture of Life Safety for the Fire Service," *Firehouse*, June 28, 2022, https://www.firehouse.com/safety-health/article/21272597/creating-a-culture-of-life-safety-for-the-fire-service-kristopher-blume.
3. Stephen M. De Lisi, "Organizational Culture in the Fire Service," *Fire Engineering*, August 1, 2005, https://www.fireengineering.com/firefighting/organizational-culture-in-the-fire-service/.
4. Kristopher Blume, "Mission-Driven Culture: A Path to Success," *Fire Engineering*, August 1, 2022, https://www.fireengineering.com/firefighting/mission-driven-culture-a-path-to-success/.
5. Ibid.
6. Ibid.
7. Ibid.
8. Dan Heath, *Upstream: The Quest to Solve Problems before They Happen* (New York: Avid Reader Press, 2020), p. 223.
9. Blume, "Mission-Driven Culture."
10. Chris Groscurth, "Why Your Company Must Be Mission-Driven," Gallup, March 6, 2014, https://www.gallup.com/workplace/236537/why-company-mission-driven.aspx.
11. Blume, "Mission-Driven Culture."
12. Ibid.
13. Ibid.
14. Mark Smith and Don Whittemore, "Mission-Driven Culture," *MC Solutions*, accessed August 10, 2023, https://www.industrialfireworld.com/532511/mission-driven-1.
15. Smith and Whittemore, "Mission-Driven Culture."
16. Blume, "Mission-Driven Culture."
17. Ibid.
18. Ibid.
19. Ibid.
20. "Brooks International's View on Leadership," Brooks International," accessed November 12, 2024, https://brooksint.com/brooks-internationals-view-leadership/.
21. Smith and Whittemore, "Mission-Driven Culture."
22. Blume, "Mission-Driven Culture."

23. Ibid.
24. Ibid.
25. Ibid.
26. Ibid.
27. Ibid.
28. Ibid.
29. Ibid.
30. Ibid.
31. Ibid.
32. "Using NFPA 1300 as a Tool to Comply with CMS Requirements for an Emergency Preparedness Program," NFPA, June 2019, https://ncw-herc.org/wp-content/uploads/2020/09/WhitePaperNFPA1300.pdf.
33. Matt Rowley, "5 Emerging Trends in the Fire Service in 2020" *The Link*, February 10, 2020, https://www.columbiasouthern.edu/blog/blog-articles/2020/february/emerging-trends-in-the-fire-service-2020/.
34. *2018 ESO Fire Trends Report* (Austin, TX: ESO Solutions, 2018).
35. "An Overview of Firefighter Accountability Systems to Improve Safety," Incident Management Technology, November 3, 2021, https://pamsfd.com/an-overview-of-firefighter-accountability-systems-to-improve-safety/.
36. Alain Rancourt, "The Success of a Good Accountability System," *Fire Engineering*, June 12, 2023, https://www.fireengineering.com/firefighting/the-success-of-a-good-accountability-system/.
37. Matt Wilson, "Culture and Accountability in the Fire Service," FireDawg.us, January 3, 2021, https://firedawg.us/commentary/leadership/culture-and-accountability-in-the-fire-service/.
38. International Association of Fire Chiefs, "Officer Development Program Leadership Development Certification Series," accessed November 11, 2024, https://www.iafc.org/learn-and-develop/officer-development-program/ODPCertification.
39. "Courses," U.S. Fire Administration, last reviewed June 23, 2023, https://www.usfa.fema.gov/nfa/courses/.
40. Ibid.
41. Frank Viscuso, *Step Up and Lead.* (Tulsa, OK: PennWell, 2013).
42. Kristopher T. Blume, "How to Change Fire Service Culture," FireRescue1, June 8, 2020, https://www.firerescue1.com/fire-department-management/articles/how-to-change-fire-service-culture-Kkv0y6FgVq2gCKls/.
43. "What Is Business Continuity?" Cisco, accessed November 12, 2024, https://www.cisco.com/c/en/us/solutions/hybrid-work/what-is-business-continuity.html.
44. Blume, "How to Change Fire Service Culture."

45. Eric Charles, "Explaining Behaviorism: Operant and Classical Conditioning," *Psychology Today*, February 28, 2014, https://www.psychologytoday.com/us/blog/fixing-psychology/201402/explaining-behaviorism-operant-classical-conditioning.
46. R. Narayan et al., "Constructivism—Constructivist Learning Theory," in *The Handbook of Educational Theories*, ed. Beverly J. Irby et al. (Charlotte, NC: Information Age, 2013), pp. 169–83, https://psycnet.apa.org/record/2013-11995-015.
47. Heath, *Upstream*, p. 10.

3. Dilemmas, Dissonance, Ethics, and Legal Concerns in the Fire Service

1. C. G. Lynch, "5 Things Carly Fiorina Has Learned about Leadership," *CIO*, May 4, 2007, https://www.cio.com/article/272146/careers-staffing-5-things-carly-fiorina-has-learned-about-leadership.html
2. Maureen Kilcullen and Judith Ohles Kooistra, "At Least Do No Harm: Sources on the Changing Role of Business Ethics and Corporate Social Responsibility," *Reference Services Review*, 27, no. 2 (June 1, 1999): 158–78, https://doi.org/10.1108/00907329910275150.
3. "What Is Corporate Social Responsibility (CSR)?" HEC Paris, accessed November 12, 2024, https://www.hec.edu/en/faculty-research/centers/society-organizations-institute/think/so-institute-executive-factsheets/what-corporate-social-responsability-csr.
4. "Local Government Fire Services—Authority and Requirements," MRSC, last modified February 23, 2024, https://mrsc.org/explore-topics/public-safety/fire-ems/fire-services-authority.
5. "Fire Service Grants and Funding," U.S. Fire Administration, accessed November 12, 2024, https://www.usfa.fema.gov/a-z/grants/index.html
6. Sarah (Wilson) Handler, "Fire and EMS Funding in the American Rescue Plan," EMS1, March 11, 2021, https://www.ems1.com/legislation-funding/articles/fire-and-ems-funding-in-the-american-rescue-plan-e2JRDH5f0P5mYv9u/.
7. "Fire Department Funding: Where It Goes, and Where It Doesn't," *Firefighters and EMS Fund*, February 20, 2020, https://www.fireandemsfund.com/fire-department-funding-where-it-goes-and-where-it-doesnt/.
8. Ibid.
9. Brian Osgood, "US Federal Firefighters Grapple with Staffing Crisis," *Al Jazeera*, July 28, 2022, https://www.aljazeera.com/news/2022/7/28/us-federal-firefighters-grapple-with-staffing-crisis.

10. Dan Heath, *Upstream: The Quest to Solve Problems before They Happen* (New York: Avid Reader Press, 2020), p. 13.

4. Managing the Brand

1. "Doctors, Military Officers, Firefighters, and Scientists Seen as Among America's Most Prestigious Occupations," *PR Newswire*, September 10, 2014, https://www.prnewswire.com/news-releases/doctors-military-officers-firefighters-and-scientists-seen-as-among-americas-most-prestigious-occupations-274579231.html.

5. The High Cost of Low-Fidelity Leadership

1. Kelly, Jack, "The Trust Gap Between Employers and Workers," *Forbes,* March 12, 2024, https://www.forbes.com/sites/jackkelly/2024/03/12/the-trust-gap-between-employers-and-workers/.
2. James L. Heskett, *The Culture Cycle: How to Shape the Unseen Force That Transforms Performance* (Upper Saddle River, NJ: FT Press, 2012).
3. Boris Groysberg and Michael Slind, "The Silent Killer of Big Companies," *Harvard Business Review*, October 25, 2012, https://hbr.org/2012/10/the-silent-killer-of-big-companies.

6. Discipline

1. "What Is The Most Important Leadership Attribute?" Robert Half, September 22, 2016. Robert Half Management Resources. https://press.roberthalf.com/2016-09-22-What-Is-The-Most-Important-Leadership-Attribute.

7. Accountability—the Antidote to Chaos

1. Dan Heath, *Upstream: The Quest to Solve Problems before They Happen* (New York: Avid Reader Press, 2020), p. 88.
2. John C. Maxwell, *Failing Forward: Turning Mistakes into Stepping-Stones for Success* (Nashville: HarperCollins Leadership, 2007).
3. Heath, *Upstream*, p. 168.
4. Maxwell, *Failing Forward.*
5. Ibid.

8. The Importance of Firefighter Health

1. Kristopher T. Blume, "Psychological Safety: Not Just Another Word for Trust in the Fire Service," FireRescue1, March 15, 2023, https://www.firerescue1.com/wellness-week/articles/psychological-safety-not-just-another-word-for-trust-in-the-fire-service-TrlPij9vC2iSf04Z/.
2. Ibid.
3. Ibid.
4. Ibid.
5. Ibid.
6. Ibid.
7. Ibid.
8. Ibid.
9. Ibid.
10. Ibid.
11. Jada Hudson, *Firefighter Emotional Wellness: How to Reconnect with Yourself and Others* (Tulsa, OK: Fire Engineering Books and Videos, 2022).
12. Blume, "Psychological Safety."
13. Ibid.
14. Ibid.
15. Ibid.
16. Brett T. Litz et al., "Moral Injury and Moral Repair in War Veterans: A Preliminary Model and Intervention Strategy," *Clinical Psychology Review*, 29, no. 8 (December 1, 2009): 695–706, https://doi.org/10.1016/j.cpr.2009.07.003.
17. Wanda J. Christie and Sara Jones, "Lateral Violence in Nursing and the Theory of the Nurse as Wounded Healer," *Online Journal of Issues in Nursing*, 19, no. 1 (December 9, 2013), https://doi.org/10.3912/ojin.vol19no01ppt01.
18. Craig J. Bryan et al., "Measuring Moral Injury: Psychometric Properties of the Moral Injury Events Scale in Two Military Samples," *Assessment*, 23, no. 5 (October 2016): 557–70, https://doi.org/10.1177/1073191115590855.

9. Five Laws of Leadership

1. *Standard on Fire Department Infection Control Program*, NFPA 1581 (Quincy, MA: National Fire Protection Association, 2022), https://www.nfpa.org/codes-and-standards/nfpa-1581-standard-development/1581.

2. Jocko Willink and Leif Babin, *Extreme Ownership: How U.S. Navy SEALs Lead and Win* (New York: St. Martin's, 2015), p. 277.

10. CT Developing Your Team Leadership Ethos

1. Kris Blume, "Creating a Motivational Fire Service Environment," *FireHouse*, June 28, 2021, https://www.firehouse.com/leadership/article/21228432/creating-a-motivational-fire-service-environment.
2. Eithne Heffernan et al., "Factors that motivate individuals to volunteer to be dispatched as first responders in the event of a medical emergency: A systematic review protocol," *HRB Open Research 2* (September 14, 2020): 34, https://doi.org/10.12688/hrbopenres.12969.2.
3. Kristopher T. Blume, "'See One, Do One, Teach One': Mentorship in the Fire Service," FireRescue1, November 16, 2021, https://www.firerescue1.com/leadership/articles/see-one-do-one-teach-one-mentorship-in-the-fire-service-NwvWLmfr2UTvbcWZ/.
4. Ibid.
5. Ron Hiraki, "Stepping Up: Mentoring," *Fire Engineering, March 12, 2010,* https://www.fireengineering.com/leadership/hiraki-mentoring/.
6. Blume, "'See One, Do One, Teach One.'"
7. Ibid.
8. Ilya Shmidt, "How to Start a Mentorship Program: A Complete Step-By-Step Guide," *Forbes*, June 2, 2022, https://www.forbes.com/councils/forbesbusinesscouncil/2022/06/02/how-to-start-a-mentorship-program-a-complete-step-by-step-guide/.
9. Ibid.
10. Blume, "'See One, Do One, Teach One.'"

11. Developing Community and Organizational Equity

1. Rosemary Hays-Thomas, *Managing Workplace Diversity and Inclusion* (New York: Routledge, 2016), https://doi.org/10.4324/9780203127049.
2. Suzanne C. Segerstrom and Gregory E. Miller, "Psychological Stress and the Human Immune System: A Meta-analytic Study of 30 Years of Inquiry," *Psychological Bulletin*, 130, no. 4 (January 1, 2004): 601–30, https://doi.org/10.1037/0033-2909.130.4.601.

3. WebMD Editorial Contributor, "How Does Mental Health Affect Physical Health?" WebMD, reviewed September 20, 2023, https://www.webmd.com/mental-health/how-does-mental-health-affect-physical-health.
4. Taylor Cox, Jr., *Creating the Multicultural Organization: A Strategy for Capturing the Power of Diversity* (San Francisco: Jossey-Bass, 2001).
5. Binford, Tammy, "Getting Social Media Recruiting on the Right Track for Diversity," *HR Daily Advisor*, April 12, 2018, https://hrdailyadvisor.blr.com/2018/04/12/getting-social-media-recruiting-right-track-diversity/.

12. Listening and Crucial Conversations

1. McChrystal, Stanley, David Silverman, Tantum Collins, and Chris Fussell, *Team of Teams: New Rules of Engagement for a Complex World*, New York: Portfolio Penguin, 2015.

13. Hiring, Interviewing, and Onboarding

1. Shanks, Roxanne, and Fitch, Jay, "EMS Providers and Leaders Share Their Perception of the State of EMS: EMS1 2019 EMS Trend Report (Part 1)," *Pulsara, May 28, 2019.* https://www.pulsara.com/blog/ems-providers-and-leaders-share-their-perception-of-the-state-of-ems-ems1-2019-ems-trend-report-part-1.
2. Kris Blume, "Command Resonance," *Firehouse*, November 15, 2021, https://www.firehouse.com/leadership/article/21241661/command-resonance.
3. Ibid.
4. Ibid.
5. Goleman, Daniel, Richard E. Boyatzis, and Annie McKee, *Primal Leadership: Realizing the Power of Emotional Intelligence*, Boston: Harvard Business School Press, 2002.
6. Ibid.
7. Ibid.
8. Frank Ricci, *Command Presence: Increase Your Influence* (Tulsa, OK: Fire Engineering Books & Videos, 2023).
9. Blume, "Command Resonance."
10. Martine, Christy Ann, and Green, John Mark, *The Light You Need: Words of Hope and Encouragement*, Independently published, December 20, 2022.
11. Ibid.

14. Stoicism, Resilience, and the Fire Service

1. Kris Blume, "Stoicism and the Art of Fighting Fire," LinkedIn, April 7, 2020, https://www.linkedin.com/pulse/stoicism-art-fighting-fire-kris-blume/.
2. Angela Duckworth, *Grit: The Power of Passion and Perseverance* (New York: Scribner, 2016).
3. Ibid.
4. Kristopher T. Blume, "Thinking Critically, Acting Decisively in the Fire Service," FireRescue1, November 2, 2020, https://www.firerescue1.com/career-1/articles/thinking-critically-acting-decisively-in-the-fire-service-FcKnFMmRuuRPC1Oc/.
5. Daniel Kahneman, *Thinking, Fast and Slow* (New York: Farrar, Straus and Giroux, 2013).
6. Malcolm Gladwell, *Outliers: The Story of Success* (New York: Little, Brown, 2008).
7. George S. Patton, "Quotes," General Patton Memorial Museum, http://generalpatton.com/index.php/quotes/.
8. Schuett, Don, "Skate to Where the Puck Is Going: Position for the Future to Achieve a Competitive Edge," *Digital Realty,* https://www.digitalrealty.com/resources/articles/skate-to-where-the-puck-is-going-position-for-the-future-to-achieve-a-competitive-edge.
9. Jane Bentson, "Could You Be More Decisive?", Jane Benston, https://janebenston.com/decisive-decision-maker/.
10. Blume, "Thinking Critically, Acting Decisively."

15. The Importance of Strategy and Trust

1. Dan Heath, *Upstream: The Quest to Solve Problems Before They Happen* (New York: Avid Reader Press, 2020), p. 92.
2. Kristopher T. Blume, "Stop Planning Strategically, Start Thinking Strategically," *Firefighter Nation*, August 4, 2022, https://www.firefighternation.com/firerescue/stop-planning-strategically-start-thinking-strategically/.
3. Ibid.
4. Ibid.

5. Ibid.
6. Ibid.
7. Heath, *Upstream, p. 16.*
8. Blume, "Stop Planning Strategically, Start Thinking Strategically."
9. Heath, *Upstream, p. 209.*
10. Blume, "Stop Planning Strategically, Start Thinking Strategically."
11. Marja Naaranoja and Asiya Kazmi, "Cultivating Strategic Thinking in Organizational Leaders by Designing a Supportive Work Environment," *Procedia - Social and Behavioral Sciences 181* (2015): 43–52, accessed *September 19, 2021,* https://www.researchgate.net/publication/277026654_Cultivating_Strategic_Thinking_in_Organizational_Leaders_by_Designing_Supportive_Work_Environment.
12. Ibid.
13. Kris Blume, "Trust: Construct and Fire Service Fundamental," *Firehouse,* March 14, 2022, https://www.firehouse.com/leadership/article/21256361/trust-construct-and-fire-service-fundamental.
14. Michelle and Dennis Reina, *Reina Trust Building,* https://reinatrustbuilding.com/.
15. Blume, "Trust."
16. Samuel A. Culbert and John J. McDonough, "The Politics of Trust and Organization Empowerment," *Public Administration Quarterly 10*, no. 2 (Summer 1986): 171–188, published by Sage Publications, Inc.
17. Ibid.
18. Ibid.

INDEX